This Land Is Your Land

Adolph and Olaus Murie at their camp on the Savage River, Mount Mckinley (Denali) National Park in 1922. Photo used with permission of the Murie Center.

This Land Is Your Land

The Story of Field Biology in America

MICHAEL J. LANNOO

The University of Chicago Press
Chicago and London

The University of Chicago Press, Chicago 60637
The University of Chicago Press, Ltd., London
© 2018 by The University of Chicago
Published 2018
Printed in the United States of America

27 26 25 24 23 22 21 20 19 18   1 2 3 4 5

ISBN-13: 978-0-226-35847-5 (cloth)
ISBN-13: 978-0-226-58089-0 (paper)
ISBN-13: 978-0-226-35850-5 (e-book)
DOI: https://doi.org/10.7208/chicago/9780226358505.001.0001

Library of Congress Cataloging-in-Publication Data

Names: Lannoo, Michael J., author.
Title: This land is your land: the story of field biology in America /
    Michael J. Lannoo.
Description: Chicago: The University of Chicago Press, 2018. |
    Includes bibliographical references and index.
Identifiers: LCCN 2017056169 | ISBN 9780226358475 (cloth: alk. paper) |
    ISBN 9780226580890 (pbk: alk. paper) | ISBN 9780226358505 (e-book)
Subjects: LCSH: Biology—Fieldwork—United States—History. |
    Biologists—United States—Biography. | Nature study—United States—History. |
    Conservation biology—United States—History.
Classification: LCC QH319. A1 L36 2018 | DDC 570.92—dc23
LC record available at https://lccn.loc.gov/2017056169

♾ This paper meets the requirements of ANSI/NISO Z39.48-1992 (Permanence of Paper).

To Frank Egerton, my mentor, and Bob Johnson, my sensei
And to Dan Townsend, James Albert, Joe Eastman, and Allan Pessier,
who invited me along

We need a literature of science which shall be readable.

THEODORE ROOSEVELT, *Biological Analogies in History* (1910)

The men [the Midwest] produced over a period of several [nineteenth-century] generations showed such family resemblance that until immigration drowned them under they constituted a strong regional type, and their virtues as exemplified in a Lincoln or a Mark Twain force the conclusion that this crude society with its vulgar and inadequate culture somehow made notable contributions to mankind.

WALLACE STEGNER, *Beyond the Hundredth Meridian* (1954)

Adam Smith needs revision. The best result will come if everyone in the group does what is best for himself AND the group.

A BEAUTIFUL MIND (2001)

The sciences, by softening the manners of men, have, perhaps, contributed more than wise and beneficial laws to the happiness of society.

JEAN-FRANÇOIS DE GALAUP, Comte de La Pérouse
(Beidleman 2006, 16)

Contents

Preamble

They stand out. You first spot the contrast between the clothes—loose and faded, stained and torn, maybe patched but probably not—and the body—weathered and fit.[1] Then you notice the face, lined from seasons of being battered by sun and wind, and scarred from a lifetime of living, some of it away from first-rate medical care. Human experiences leave their marks, like inscriptions, and, as with any difference, you notice.

E. O. Wilson observed that field biologists have a lot more "gee whiz" or "sense of wonder" than other kinds of scientists.[2] Field biologists are people who make a life and a living out of heeding Edward Abbey's call to "ramble out yonder and explore the forests, climb the mountains, bag the peaks, run the rivers, [and] breathe deep of that yet sweet and lucid air."[3] Clearly they enjoy themselves—just listen to the ease and character of their laughter[4]—the epitome of the notion that in wilderness lays human wellness. These are folks, as Abbey advocated, who have kept their brains in their heads and their heads firmly attached to their bodies; bodies obviously attuned and active. Joseph Campbell summed up this approach to life in a one-liner from his masterwork, *The Power of Myth*: "What we're all really seeking . . . is an experience where we can feel the rapture of being alive."[5]

Campbell's West Coast friends Ed Ricketts and John Steinbeck were referring to this "aliveness" when they wrote in their collaboration, *Sea of Cortez*:

> We sat on a crate of oranges and thought what good men most biologists are, the tenors of the scientific world—temperamental, moody, lecherous, loud laughing, and healthy. . . . Your true biologist will sing you a song as loud and as off-key as will a blacksmith, for he knows that morals are too often diagnostic of prostatitis and stomach ulcers. Sometimes he may proliferate a little too much in all directions, but he is as easy to kill as any other organism, and

meanwhile he is very good company, and at least he does not confuse a low
hormone productivity with moral ethics.[6]

There is power here. A field biologist is the sort of person many people
want to be—the image now being sold by tony outdoor clothing companies.
In January 1998, Joe Eastman and I were sitting at a small table in the Quonset
hut bar misnamed the Coffee House at the U.S. Antarctic base at McMurdo.
We had just packed and secured our nearly one thousand ice fish specimens
collected during a three-week zigzag cruise over the Ross Sea on the National
Science Foundation–leased icebreaker *Nathaniel B. Palmer*. Lying low, we were
sheltering from a pop-up blizzard, enjoying a few quiet, reflective beers, when
the door burst open and a small group of graduate students stomped in. They
had spent much of the austral summer exploring the Mars-like features of Antarctica's inland dry valleys and had just returned. A short time later more students arrived, then more, as one lost acquaintance after another was renewed.
The noise level doubled and doubled again as the booze flowed. These kids
had just been through the field experience of a lifetime at the one place on
Earth that could provide it, and they knew it. Many understood this was
going to be the last time they would be in Antarctica. They knew they had
been changed, and that such transitions must be celebrated. Joe and I left at
midnight with the party still roaring but the blizzard abated. We pushed open
the heavy door and were met with calm air, blue skies, and snow-blind sunlight. The noisy C-130s were landing, and our names were on the manifest
on the first flight out to Christchurch.

Part of the euphoria of a recently completed field season, especially in a
remote area of the planet, is simply surviving it. Antarctica is uninhabitable
by unsupplied humans, and in Antarctica reminders of death are everywhere.
Scott's Hut sits on a little promontory just north of the McMurdo Station. The
one unassailable truth about the Heroic Age of Antarctic Exploration—the
period of time when Robert Falcon Scott and Ernest Shackleton were leading
expeditions south toward the Pole—was if you made it back to Scott's Hut,
you lived; if you did not, you died. Life was simpler in those days. And you
are reminded of that when Scott's Hut is the first object that registers in your
befuddled mind after you leave a bar called the Coffee House at midnight
in bright sunshine, a few hours before the U.S. Navy is scheduled to fly you
home.[7]

Not all field biologists survive their field seasons. On August 3, 1993, the
light plane carrying Ted Parker, Al Gentry, and five others crashed into a
mountainside in western Ecuador.[8] It was supposed to be a routine flight, to
conduct a rapid assessment of the plants and animals occupying the remaining

rainforest in the southwestern portion of that country. No flight plan had been filed, navigational errors were made, and the mountain was hidden in a cloudbank. Soon after the crash, the pilot and Eduardo Aspiazu, an Ecuadorian conservation biologist working with Fundación Natura, died. Parker's fiancé, Jacqueline Goerck, suffered serious injuries to her vertebrae and ankle. Ignoring the pain, she and Ecuadorian biologist Carmen Bonifaz worked their way down the thickly forested mountainside to find help. That night, Al Gentry died. By the time rescuers arrived the next afternoon, Ted Parker had also died. The Ecuadorian biologist Alfredo Luna was alive, barely; he was evacuated and eventually recovered. At the time of their deaths, Parker was the preeminent neotropical ornithologist, Gentry the preeminent neotropical botanist. Parker had committed over four thousand birds to memory; Gentry could identify over six thousand species of South American woody plants by sight or smell. There was nobody on Earth better at what they did than Ted Parker and Al Gentry; there is still nobody better.[9]

Even the casually interested observer will have noted the recent tendency to romanticize field biologists. Already legends within their fields, Parker's and Gentry's deaths propelled them to immortality across the disciplines that comprise conservation biology. In eulogizing them, and championing the efforts of their sponsoring Rapid Assessment Program, Russell Mittermeier, president of Conservation International, wrote: "The romantic adventure-filled nature of [this] work struck a chord with a wide audience—these after all, were the true 'Indiana Joneses' of the world."[10]

Because the fictional Indiana Jones was an archaeologist, this comparison seems strange, but in a twisted way it isn't. With his wide-brimmed fedora, leather bomber jacket, chinos, and sidearm, the image of Indiana Jones has been co-opted—through apparent unspoken general agreement—to be the "ideal" of a modern field biologist: swashbuckling, no-nonsense, often in mortal danger, and usually getting the girl. This Indiana Jones iconography has legs and appears elsewhere. Tim Gallagher has spent much of the last decade searching for the great lost woodpeckers of the New World: first the U.S./ Cuban ivory-billed woodpecker, more recently Mexico's magnificent imperial woodpecker. In praising Gallagher's latest book, Stephen Bodio described it as "a blend of natural and tragic human history and Indiana Jones–style adventure."[11] And in complimenting Samantha Weinberg's book *A Fish Caught in Time*, about the search for the prehistoric coelacanth, the London *Mail on Sunday* commented: "Reads like some classic Spielberg creation—Indiana Jones let loose in a real-life Jurassic Park."[12] The most incongruous Indiana Jones comparison comes from Zachary Jack, editor of a recent compilation of Liberty Hyde Bailey's work, who wrote, in a hyperbole reflecting the current

popular standard: "Bailey would become for Cornell faculty, staff, and students what the fictional professor Indiana Jones would become for his [Uni-
versity of Chicago] archeology students: a jaw-dropping wonder."[13]

Mythology aside, field biology mostly involves a lot of hard work—late
nights skinning birds or pinning insects, meticulous field notes written by
the dim light of a headlamp or maybe a full moon: Olaus Murie would stay
up past midnight recording his day;[14] in the Himalayas, George Schaller spent
up to three hours each evening transcribing records with a ballpoint pen he had
to heat over a candle to keep the ink flowing;[15] Al Gentry's frenetic pace meant
he wrote few field notes at all.[16] In considering this effort and inconvenience,
not everyone agrees with the notion of field biologists as Indiana Jones–style
adventurers. As John Wesley Powell said about his Colorado River expeditions:
"The exploration was not made for adventure, but purely for scientific purposes, geographic and geologic."[17]

Following this sentiment, Steinbeck and Ricketts declared:

> . . . the atavistic urge toward danger persists and its satisfaction is called ad-
> venture. . . . We had no urge toward adventure. We planned to collect marine
> animals in a remote place on certain days and at certain hours indicated on
> the tide charts. To do this we had, in so far as we were able, to avoid adventure.
> Our plans, supplies, and equipment had to be more, not less, than adequate;
> and none of us was possessed of the curious boredom within ourselves which
> makes adventurers or bridge players.[18]

As much as field biologists might wish to avoid it, at times adventure
comes to them. On Christmas Eve 1993, I was with a small group that included the ichthyologists James Albert and Tom DiBenedetto. We had just
spent a long day collecting several species of electric fish and were sponging
off and kicking back in the black waters of the Atabapo River just south of
the Venezuelan town of San Fernando de Atabapo. The Atabapo flows north
through balsa wood forests and is therefore a black water river—rich with
organic tannins. At San Fernando, the Atabapo joins the white limey water
of the Guaviare River before continuing on to flow into the Orinoco a few
miles north. We had been working the Atabapo above its confluence with the
Guaviare because mosquito larvae do not survive its acidic organic waters. If
you can avoid mosquitoes, you can prevent malaria. None of us wanted that.

We were a long way from home, the sun was setting, and our families back
in the States were opening presents and running through familiar holiday
routines. As we wound down, our thoughts naturally turned toward home,
and we began feeling a collective sinking melancholy. Then, just after we got
quiet, the water began roiling as a pod of pink dolphins showed up—playful,

excited, swimming among us, circling, surfacing, and blowing, their strange, finless backs so close we could touch them. Our glumness instantly evaporated on this bit of serendipitous adventure.

Adventure or not, in its most general sense, field biology is simply studying nature in nature, which could encompass everything from George B. Schaller's grand Karakorum and Himalayan expeditions to investigate the behaviors of snow leopards and their ungulate prey,[19] to a humble midafternoon backyard teatime break to observe hummingbirds working domesticated trumpet flowers. It is the former rather than the latter approach I'll be emphasizing here. And so, to this general definition, I will add that because primary observations must come first, field biology will usually involve some sort of lifestyle disruption due to the need to be at that special place during that particular time when one can best, or perhaps only, attend to that specific subject. In short, field biology involves some level of self-sacrifice. There is a cost—travel, sleep, comfort, perhaps relationships. This tradition of self-sacrifice goes back a long way. In 1834, when Thomas Nuttall and John Kirk Townsend were collecting specimens of never-before-described plants, birds, and mammals during the Wyeth Expedition to the Oregon country, they placed the safety of their collections over personal comfort and well-being. Shortly after beginning their expedition, Nuttall wrote, "Already we have cast away all our useless and superfluous clothing and have been content to mortify our natural pride, to make room for our specimens."[20]

There is a cumulative effect of these experiences that you notice in the ragged appearance and self-assured manner of field biologists. To them, such a disjointed lifestyle is never a sacrifice. To them, fieldwork is both the means and the end to a life well-lived and thoroughly tested. (George Schaller described it as the "kind of satisfaction that only effort and endurance can provide."[21]) When I stay up all night on a pocket prairie in southern Indiana checking seasonal wetlands for breeding crawfish frogs, I record capture times in my Rite in the Rain notebook but never consciously register the clock until I hear the 5:50 AM traffic (~six cars) signaling the shift change at the coal mine a mile north. At that point, I know sunrise is near and my night almost over.

George Kruck Cherrie recounted a similar early twentieth-century nighttime collection experience at Campo Santo, Brazil:

My attention was fixed on the adobe wall in front of me. Rays from my powerful lantern illuminated a white disk on the wall fully ten feet in diameter. Between the lantern and the disk, a distance of from fifteen to twenty feet, was a cone of light sharply defined against the blackness of the night. Within a few seconds this cone became populated with hundreds of flying, buzzing, circling, darting insects. . . . Only when a beautiful big moth circled lazily into the light and his wing-spread was shadowed large against the white wall behind him, did I make a wide sweep with my net and begin the real work of the evening. . . . In my enthusiasm I felt as if I could go on swinging my net all night long. . . . Never in my life had I spent so riotous a time at collecting. When fatigue did come it came with a rush. I had lost all account of time . . . [and] felt it was the most successful evening's work with insects I had ever spent. Having extinguished my lantern, I made my way slowly back to my lodgings and dropped contentedly into my hammock.[22]

The golden age of field biology in North America lasted from the last half of the nineteenth century until perhaps just after the Second World War. During these years, transportation in the form of railroads and horse-drawn wagons was, for the first time, available, then modernized as roads were built and automobiles became common. (Annie Alexander, collecting for her Museum of Vertebrate Zoology, named her six-wheeled, tracked Ford "Blundie" and her Franklin roadster "Birdie."[23]) Natural history surveys were organized, colossal museums constructed to house their specimens, and field stations cobbled together to civilize the experience.

There was geographical bias. Many of the finest field biologists in history came out of the U.S. Midwest in the nineteenth century.[24] They grew up at a time when the Midwest was frontier; when hunting, fishing, and trapping were a part of a boy's life, and to be successful you had to know the habits and habitats of the animals you sought. Many of these early biologists ended up on the East Coast, working for big institutions. They were considered "rough," "independent," and "naïve" by eastern standards. Their center was the Cosmos Club, in Washington, DC, founded in 1878 by the Illinois-raised John Wesley Powell and his colleagues. There, these men (nearly almost always men in those early days) could read and discuss issues among themselves in a more relaxed setting, with no eavesdropping.

Most histories of this period favor the big eastern institutions driving such explorations, such as the Philadelphia Academy of Natural Sciences, the American Museum of Natural History, the Smithsonian Institution, Harvard's Museum of Comparative Zoology, the New York Botanical Garden, and the institutions at Woods Hole.

Historians of biology are quick to pronounce that in the early twentieth century, lab biology arose from, and replaced, field biology. But that is like saying because reptiles, through amphibian intermediates, evolved from fishes, reptiles replaced fishes. Not true, as every sportsman knows. Fishes are still very much with us, and in fact today outnumber reptiles almost 3 to 1 (~27,300 fish species[25] compared with 10,272 reptile species[26]). They also taste better.

Field biology continued to thrive until the mid-1950s, when the new and then contrary field of molecular biology—which exploded in popularity following James Watson and Francis Crick's discovery of DNA and its method of replication—attracted many of that generation's most talented young biologists. Richard Bovbjerg, the former director of the Iowa Lakeside Laboratory, commented on one consequence of this rebuff—field station attendance: "Across the land there has been a significant drop in registration at field stations. Some stations have gone under. The average registration drop over 10 years has been 40%."[27] This was not just the case at field stations; it became a societal phenomenon. In 1956 attendance at the Missouri Botanical Garden, which would later come to host Al Gentry, reached an all-time low.[28]

Field biologists also had to confront the contingent nature of their science. As Robert Kohler pointed out, run a laboratory benchtop experiment anywhere in the world and as long as all of the variables are controlled, the results obtained should be the same—ideas can be tested and retested, and variables can be manipulated to determine effect.[29] Not so with natural history. Most field biologists conducting long-term research studies will tell you that no two years are ever the same—that some years remind them of others but the fine details differ. Mark Twain is credited with saying: "History doesn't repeat itself, but it does rhyme."[30] In the midst of this mass variation, generalizations come slowly and are usually conditional. Viewed in this light, it becomes easy for bone-weary field biologists to consider themselves mere stamp collectors; that is, until they realize the greatest theory in the history of biology—Charles Darwin's and Alfred Russel Wallace's theory of evolution through natural selection—emerged from the very same techniques and thought processes employed by today's field biologists.[31]

Being aware of these pitfalls, Paul Dayton defended natural historians: "Natural history is the foundation of ecology and evolution science. There is no ecology, no understanding of the function of ecosystems and communities, no restoration, or in fact, little useful environmental science without an understanding of the basic relationships between species and their environment, which is to be discovered in natural history."[32]

Today, field biology is enjoying a resurgence,[33] due in part to the efforts of E. O. Wilson, Paul Dayton, and other superbly talented scientists such as Harry Greene, Marty Crump, and Rafe Sagarin, who have popularized their fieldwork.[34] Field station attendance is up, and books such as *Naturalist* and *The Essential Naturalist* have been published.[35] This reawakening is also due to several other factors, including the recognition that ecological relationships are complicated—more complicated than even our most sophisticated computer-generated statistical/mathematical models. Modeling life is a lot like living it: doing it well is more difficult than you might at first think. Rather than having such models be the last word on the subject, which we as younglings were taught to believe, ecologists are now recognizing, in the words of George E. P. Box, that "all models are lies; some of them are useful."[36]

And finally, this reawakening of field biology must in part be due—no one can deny it—to the Indiana Jones swag factor: the observation that field biologists have a charismatic way of carrying themselves, one that cannot possibly be derived from a nine-to-five lifestyle spent in front of a computer screen or a DNA sequencer, and must instead come from the confidence—so apparent in late eighteenth- and early nineteenth-century field biologists—of being comfortable and capable outdoors.

Having now just emerged from a bottleneck, field biology now needs to explore its origins and claim its history. As with every stenosis, selection has occurred. Clean lines of descent have blurred as some lineages have gone extinct, others have hybridized, and still others have gone through such severe selection pressures that they have transformed and are now nearly unrecognizable. Because of this, modern field biologists find themselves asking some fundamental existential questions, such as where did we come from, do we have a cohesive story we can tell, and do we have a legacy?

They do. The legacy of American field biologists forms the underpinning of what Wallace Stegner called "America's best idea."[37] Here, I offer a history of field biology in North America and what it meant to the world. It will not be much about statistical or mathematical models or famous armchair biologists. Nor will it be a celebration of biology as a top-down power struggle for institutional or personal prestige. It will instead be the flipside—about bottom-up, field-based, rubber-boot natural history. Woody Guthrie wrote his songs from a "man on a box car's perspective"—as a John Steinbeck–like observer going through the world.[38] And that's how the following chapters will play out, as a lifestyle conducted by some of its most talented early practitioners—a disproportionate number of whom happened to be midwesterners.

Acknowledgments

It is appropriate that the University of Chicago Press publish this work, for the University of Chicago was at the center of many of the happenings I cover here. In essence, the University of Chicago was a private land-grant institution, coming along late, snagging talent from East Coast universities, and co-leading field biology into the age of modern quantitative ecology. I thank my editor, Christie Henry, and her staff for their nurturing. I thank Genevieve Arlie, Neil Bernstein, Holly Carver, Marty Crump, Paul Dayton, Steve Dunsky, Mark Edlund, Frank Egerton, Harry Greene, Susan Lannoo, Curt Meine, Mike Mossman, Erin Muths, Linda Rozumalski, Bill Souder, Fred Swanson, and Linda Weir for comments on earlier manuscript drafts. I thank Rochelle Stiles and Alisa Gallant for preparing the figures. Gathering the historical photographs presented an unexpected challenge, and I thank Ellen Alers, John Boardman, Randy Bovbjerg, Victor Bovbjerg, David Brakke, Susan Braxton, Julia Buckley, Joshua Caster, Christine Colburn, Heather Cole, Sheri Dolfen, Larry Dorr, Steve Dunsky, Jonathan Eaker, Fred Errington, Christina Fidler, Chris Filstrup, Doris Hardy, Alfred Gardner, Rose Gulledge, Maria Kopecky, Michael Lange, Sarah Lathrop, Mark Madison, Lisa Marine, Daniel Meyer, Jennifer Mui, Erin Muths, Katie Nichols, Keiko Nishimoto, Matthew Perry, Desirce Ramirez, Nancy Ricketts, Marguerite Roby, Rose Rodriguez, David Rumsey, Amanda Shilling, Adrienne Sponberg, Linda Stahnke, Richard Stamm, Rochelle Stiles, Heidi Stover, Rebekah Tabah, Tony Thompson, John Waggener, Katherine Walter, Linda Weir, and Kelsey Zehner. And finally, Mark Madison at the National Conservation Training Center and David Miller at the Smithsonian kindly tolerated my questions about .32-caliber shotgun bore inserts, and I thank them for their insight.

Introduction

In their collaboration, *Sea of Cortez*, John Steinbeck and Ed Ricketts wrote: "The design of a book is the pattern of reality controlled and shaped by the mind of the writer. This is completely understood about poetry or fiction, but it is too seldom realized about books of fact."[1] The challenge then, for a writer of books of fact, is discovering the pattern of reality. Enter Norman Maclean[2]:

> . . . every now and then [life] becomes literature—not for long, of course, but long enough to be what we best remember, and often enough that what we eventually come to mean by life are those moments when life, instead of going sideways, backwards, forward, or nowhere at all, lines out straight, tense and inevitable, with a complication, climax, and, given some luck, a purgation, as if life had been made and not happened.[3]

Field biology—or the process of discovering life and its interactions, in part to understand, save, and rebuild as much of it as possible—has had its share of going sideways and backwards, or nowhere at all. So has my mind, in considering field biology's patterns and processes. To tell this story as if life had been made and not just happened, I looked for fundamentals. And, as almost always happens, when you search far and wide, you eventually discover the answer right under your nose. I found mine in an old quote (mentioned in the "Preamble") by my friend Paul Dayton:

> There is no ecology, no understanding of the function of ecosystems and communities, no restoration, or in fact, little useful environmental science without an understanding of the basic relationships between species and their environment, which is to be discovered in natural history.[4]

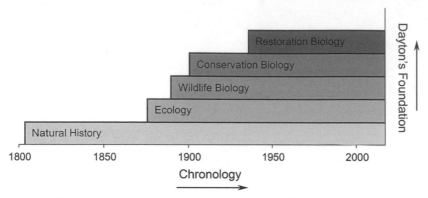

FIGURE 1. The chronological appearance of disciplines that rely on field biologists reflects Paul Dayton's insight into their relationships with each other. Note that the earliest disciplines support their successors.

These words speak for themselves and stand alone. But, at the risk of sounding metaphysical, there is something deeper in them, more implied than stated—a form of ontogeny recapitulates phylogeny. The relationships that Dayton observes today between natural history, ecology, conservation, and restoration biology (their phylogeny) are also represented in the sequence of appearance of these disciplines in the development of human environmental thought (their ontogeny; see fig. 1).[5] The metaphor here is a historic city whose modern foundations are formed by the solid stones of its original buildings.

Just as natural history preceded and today underpins ecology, natural history and ecology preceded, and today underpin, the field of conservation biology. And where important natural systems could not be preserved, natural history, ecology, and conservation biology preceded, and now underpin, the field of restoration biology. My only alteration to this logic will be to insert into this sequence the field of wildlife biology.[6] Historically, wildlife biology bridged the fields of ecology and conservation biology. It didn't have to, but history played out this way, as follows.

In North America, organized field biology originated with the pioneer *natural historians* of the nineteenth century, a period initiated by Lewis and Clark's Voyage of Discovery.[7] The task of these men was to find what was "out there" or, in modern terminology, to assess biodiversity. These earliest naturalists included Thomas Nuttall, Edwin James, and Joseph Nicollet. They were imports to western North America. A handful were Americans trained in the eastern United States; Nuttall was British, and Nicollet was French. They were the first to attempt to formally describe our species of plants and animals and where they lived—the components of North American natural diversity in the unsettled West, which at that time meant the territory beyond the Appalachians. As Alexander von Humboldt had before them, these naturalists saw and described things no European had ever seen. Their history has been well documented, although not always appreciated.[8]

The next generation of field biologists was homegrown—this land was their land—and arose just as land-grant colleges became established. These are the first men I consider in detail. They represented newly settled midwestern talent that either headed east as professionals to populate the big institutions or stayed at home as collectors for these big institutions.[9] Members of this group worked for people such as Asa Gray at Harvard, Spencer Baird at the Smithsonian, or C. Hart Merriam at the U.S. Biological Survey. They included Ferdinand V. Hayden, John Wesley Powell, and William Temple Hornaday. John Muir was in a sense a part of this group, although in characteristic fashion, while everyone else went east, Muir went west.

As with Europeans and easterners, these midwestern field men were natural historians, describing the areas they explored, collecting the species they found, then sending their field notes and specimens back east. But in contrast to their predecessors, they were describing extensions of their backyards, places they knew well. On the one hand, they had better technology, including transportation and instrumentation; on the other, they were revisiting old ground—there was almost nowhere they could go where a European had not been before. Their descriptions were more polished, and the world they came to describe was, by and large, the world we now know.

The tasks performed by field biologists

What's out there? (= Natural History [Biodiversity])
How does it fit together? (= Ecology)
How can we save the parts we want? (= Wildlife Biology)
How can we save it all? (= Conservation Biology)
How can we bring back what we could not save? (= Restoration Biology)

Timeline for the chronological appearance of disciplines

Natural History	Meriwether Lewis and William Clark	1803
Ecology	Stephen Forbes	~1875
Wildlife Biology	William Temple Hornaday	1889
Conservation Biology	Gifford Pinchot/John Muir	~1900
Restoration Biology	Norman Fassett/Aldo Leopold	~1935

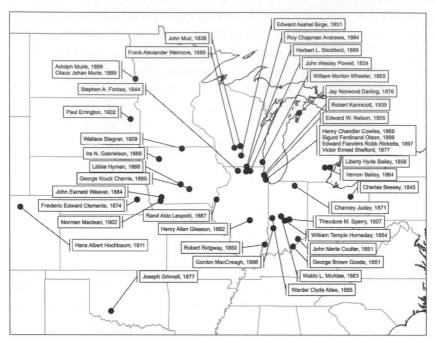

FIGURE 2. Field biologists of the Midwest, with the date and location of their births or their longtime residencies.

The efforts of these field biologists took place when Alexander von Humboldt, now largely forgotten,[10] was the world's most influential scientist.[11] Humboldt's last book, *Cosmos*,[12] may have been the inspiration for the name of the Washington, DC, scientists' club founded by John Wesley Powell and his colleagues.[13] These naturalists stocked our great museums and botanical gardens, and dominated field biology during the first three-quarters of the nineteenth century. They did their job so well that by 1878 it was all but completed. Following the great western surveys of the 1850s and 1870s,[14] the natural history of North America was largely known, although even today, in places such as the canopies of giant redwoods, we find new species or new forms of recognized species.[15] As with Ted Parker and Al Gentry, today's natural historians often work in remote areas, discovering new species in the far corners of a round world.

The second task of field biologists in North America was to figure out how all those species fit together. Specifically, as the diversity of species began to be understood, field biologists got curious about the relationships of these species to each other—how they interacted. These field biologists became our *ecologists*. While the early naturalists passively described nature and collected specimens, ecologists were aggressively quantitative. They conducted experiments to answer questions—for example, relating to cause and effect—and performed dissections of stomachs or sifted through scat to untangle food webs.[16] Aldo Leopold clearly recognized the boundaries of this transition. After interviewing an older candidate to supervise the prairie restoration at the University of Wisconsin's Arboretum, Leopold wrote:

> He is a kindly elderly gentleman of rather wide experience in horticulture and soils, with a good botanical background, but no ecology, since there was no such science in his day. . . . As a test of his command of ecological science: he had never heard of a quadrat.[17]

Many of these early ecologists trained in the Midwest and either stayed there or carried the Midwest with them as they progressed in their careers, mostly through positions opened in the newly formed land-grant institutions. This group first grew then dominated early ecological thought. It included the plant ecologists Charles Bessey, Frederic Clements, Henry Chandler Cowles, and Henry Gleason. As well, it included the animal ecologists and behaviorists Stephen A. Forbes, Victor Shelford, and Warder Clyde Allee. It also included the limnologists Edward Birge and Chancey Juday. The Ecological Society of America, cofounded by several of these men, was established in 1915.[18]

The third task of field biologists in North America became how to save the parts of nature that society found useful. That is, once species were known and their ecological relationships began to be understood, the question became how "useful" species might be conserved for the betterment of man (although not always for the betterment of the species themselves). It was from this perspective that the third type of field biologist, the *wildlife biologist*, was born. While today's progressive wildlife biologist is concerned with the welfare of all wildlife and the ecosystems that support this life, early wildlife biologists were much narrower in their focus—largely concerned with ensuring adequate numbers of game and fish (i.e., animals useful to man). Indiana-born William Temple Hornaday's nineteenth-century work on bison conservation was at the forefront of early wildlife biology, although Iowa-born Aldo Leopold rightfully can be said to have formalized the discipline with his 1932 book, *Game Management*. Practically minded ecologists became wildlife biologists. This group included the midwesterners Herbert L. Stoddard, Olaus J. Murie and his half-brother Adolph Murie, Paul Errington, Aldo Leopold, and (loosely) Jay Norwood "Ding" Darling.

The fourth task of field biologists in North America became how to save all parts of nature, not just the portions useful to man, and this idea spawned the field of *conservation biology*. It arose in fits and starts, then got traction around 1900, when an influential portion of American society—represented best, although not exclusively, by Teddy Roosevelt's Progressive Party—became disturbed by the expansive clear-cutting of the nation's forests, the near extinction of the impossibly abundant bison, the actual extinction of the impossibly abundant passenger pigeon, and the appalling slaughter of wading birds for the plume trade. Influenced by an amalgam of the Transcendentalist idea that because God is nature, nature must be preserved (promoted by John Muir) and, to the Victorian mind, the opposing idea of wise use of natural resources for man's improvement (promoted by Gifford Pinchot), the conservation movement began to gain traction. With his land ethic—"A thing is right when it tends to preserve the integrity, stability, and beauty of the biotic community. It is wrong when it tends otherwise"—Leopold blended ecology ("integrity"), the growing science of conservation ("stability"), and the romanticism of American Transcendentalists ("beauty") to construct conservation biology's philosophical underpinnings.

The fifth and final task of field biologists in North America became how to reassemble important ecosystems that had been destroyed.[19] The origins of *restoration biology* lay in the late nineteenth-century forestry practice of replanting trees in clear-cuts and include Hornaday's efforts to captive-breed bison. But organized restoration biology formally began in the mid-1930s, again in the Midwest, when Norman Fassett arranged to build Curtis Prairie on the University of Wisconsin campus. From there, the idea of ecological restoration spread fifty or so miles north, to Leopold's Shack, and from its Shack locus, the idea went global.

There is a parallel group of midwestern field biologists—the writers—who grew up surrounded by biology and who would have been top-notch field biologists themselves had they chosen this career. They cannot be left out. This group consists of the Iowans Norman Maclean and Wallace Stegner. The cartoonist and once head of the U.S. Biological Survey, Ding Darling, also belongs here. They popularized our natural history and, in turn, our ecology, wildlife biology, conservation biology, and restoration biology. They are the historians of our natural history.

This will be all of these individuals' story. But it will not just be their story—not just their history. Each of these field biologists was developing their skills against the backdrop of a frontier landscape in a country that did not yet know who it was, but was eager to find out. They contributed to their new country by discovering and interpreting its nature, in the most literal sense. While their earliest motives were centered on future commercial exploitation, over time these men developed a view of this country that ran contrary to the ideals of pure capitalism. These men loved their country for what it was—its ability to fascinate and inspire awe—rather than for the über-capitalistic notion of how much money it could make for them. Out of this perspective came "America's best idea"—national parks—which became the first in a series of "best ideas" that include national forests, national wildlife refuges, wilderness areas, and the federal legislation that underpins them. These best ideas were so good, they were exported and are now represented globally. Here, I carry the stories and insights of these field biologists forward in a trajectory that takes them from their time, through our time, and into the future. Looking backward along this trajectory, the story of field biology becomes not just a chronicle of what happened, but also what it meant—not just a narrative of history but, more critically for the world's future, the creation of an enduring legacy. The world is a different and great deal better place because of field biologists.

The Foundation

Field biologists did not arise spontaneously. Their roots lie deep in the outdoor competence of Native Americans, European explorers, mountain men, and voyageurs. As the children of pioneers who settled the Midwest, they grew up surrounded by the nature they would come to study. Their influences were not just the scientists, but also the philosophers who were trying to make sense of what the United States was and what it could become. When opportunities arose after the big eastern institutions and federal agencies were established, these men took advantage and became laborers. Later, the quality of their work propelled them into leadership positions. It happened something like this.

The Explorers

American field biologists have a long history. On April 30, 1803, the United States purchased the vast Louisiana Territory from France. As Howard Evans wrote: "The Louisiana Purchase . . . approximately doubled the size of the United States, adding 800,000 square miles of ignorance—land that had never been well explored or adequately mapped."[1] Two months later, on June 20, President Thomas Jefferson gave the following instructions to Captains Meriwether Lewis and William Clark:

> The object of your mission is to explore the Missouri river, & such principle stream of it, as, by its course & communication with the waters of the Pacific Ocean, may offer the most direct & practicable water communication across this continent, for the purposes of commerce.[2]

By the time of Lewis and Clark's Voyage of Discovery (Evans's opinion notwithstanding), large parts of western North America were already well known to Europeans. By 1800 Spanish conquistadors had a 250-year-old working knowledge of the American Southwest and the Pacific coast,[3] and Hudson's Bay Company had a 130-year history of harvesting beaver furs across the immense expanse of northern and western Canada and into the large northern West Coast region then called Oregon.

The Spanish arrived first, and their discoveries were significant. In May 1540, Hernando de Alarcón became the first European to explore the Colorado River, which he named Buena Guía.[4] Later that year, on September 8, Hernando de Alvarado discovered the Rio Grande, which he termed Río de Nuestra Señora.[5] Also in 1540, García López de Cárdenas became the first European to view the Grand Canyon.[6] A year later, in June, Hernando de Soto's men became the first Europeans to view the Mississippi River, which they called Río del Espíritu Santo,[7] from a site south of present-day Memphis. A year after, in May 1542, a group of de Soto's men, led by Luis de Moscoso Alvarado, became the first Europeans to travel down the Mississippi, from its junction with the Arkansas River to its mouth. The journey took seventeen days, and Native Americans constantly harassed the Spaniards. They found oil in Texas.[8] Also in 1542, Juan Rodríguez Cabrillo sailed up the West Coast, perhaps as far as Monterey Bay, before being driven back south by storms. Cabrillo died somewhere near Santa Barbara, but his expedition established Spain's right to the central California coast,[9] claims later reinforced by Sebastian Vizcaíno's expedition of 1602.[10] By 1650 Spaniards were so familiar with the arid Southwest and southern West Coast, they could travel throughout this region at will, relying on the knowledge of Native Americans for locations of watering holes and retreats.[11] José Mariano Mociño was the first botanist to travel extensively, from Mexico north into Canada and south into Central America. Using modern Linnaean taxonomy, Mociño is said to have described over two thousand species of plants, with a special interest in species used by the Aztecs for medicinal purposes.[12]

In 1521 Hernán Cortés and his men first saw bison, which were included in an assemblage maintained by the Aztec king Montezuma. In 1724 the historian Antonio de Solis described this early zoo:

> In the second Square . . . were the Wild Beasts, which were either presents to Montezuma, or taken by his Hunters, in strong Cages of Timber, rang'd in good Order, and under Cover: Lions, Tygers, Bears, and all others of the savage Kind which New-Spain produced; among which the greatest Rarity was the Mexican Bull; a wonderful composition of divers Animals. It has crooked Shoulders, with a Bunch on its Back like a Camel; its Flanks dry, its Tail large, and its Neck cover'd with Hair like a Lion. It is cloven footed, its Head armored like that of a Bull, which it resembles in Fierceness, with no less strength and Agility.[13]

In 1542 Francisco Vázquez de Coronado wrote: "The first time we encountered the buffalo, all the horses took to flight on seeing them, for they are horrible to the sight." Then he commented on their social organization: "We were much surprised at sometimes meeting innumerable herds of bulls without a single cow, and other herds of cows without bulls."[14]

At the same time that the conquistadors were performing their bloody brand of exploration in what would become the American Southwest, in the North, the French and British were using Hudson's Bay, as well as the St. Lawrence River and the five Great Lakes, to penetrate the North American interior. Jacques Cartier's voyages of 1534–36 took him up the St. Lawrence as far as present-day Montreal.[15] Cartier described about forty plant and sixty animal species.[16] On July 1, 1534, Cartier became the first European to record passenger pigeons (*Ectopistes migratorius*). His men were exploring the west coast of Prince Edward Island, and, as Cartier wrote: "We landed that day in four places to see the trees which are wonderfully beautiful and very fragrant. . . . The soil where there are no trees is also very rich and is covered with peas, white and red gooseberry bushes, strawberries, raspberries, and wild oats like rye. . . . There are many turtle doves, wood-pigeons, and other birds."[17] In the sixteenth century, only two species of pigeon inhabited the Maritimes; Cartier's "turtle dove" is probably the mourning dove; the "wood pigeon," the passenger pigeon.[18]

The historian John Bakeless noted, "One of the amusing things about French explorers is how very French they were. . . . [T]hey never missed good food, the prospects of good food, or a chance to make wine."[19] In the early 1600s, Samuel de Champlain established an outpost on the site of present-day Quebec City, and in 1610 he sent Etienne Brûlé west to explore the Great Lakes. Brûlé saw four of the five Great Lakes (he missed Lake Michigan) and may have been the first European to see Lake Superior.[20] French explorers and trappers followed Champlain and Brûlé and flowed into the Great Lakes region. The map of the upper Midwest began to fill in following the expeditions of Jean Nicolet (in the mid-1630s) north and west of the Ohio River and along Lake Michigan north to Green Bay, the journey of the Jesuits Isaac Jogues and Charles Raymbaut to Sault Sainte Marie (1641), and the exploration of Claude-Jean Allouez to Lake Superior and the Wisconsin interior (1669–70).[21] In May 1673 Louis Jolliet and Jacques Marquette set out along the northwest shore of Lake Michigan to Green Bay, then along the Fox River. Following a portage (at, of course, Portage, Wisconsin, near where John Muir would later spend his boyhood, and following a road that may have taken them past the future site of Aldo Leopold's Shack), they then paddled down the Wisconsin River to the "Messi-Sipi" River at Prairie du Chien, and descended the Big Muddy, stopping on July 17 south of St. Louis, just north of the Spanish border along the Arkansas River.[22] In 1682 another French explorer, René-Robert Cavelier, Sieur de La Salle, became the first European to descend the length of the Mississippi to the Gulf of Mexico.

The English foothold in the land that would become the United States was centered along the Atlantic Coastal Plain and Piedmont.[23] In 1584 Sir Walter Raleigh sailed from the West Indies north along the East Coast up the Chesapeake Bay, representing the first real English interest in America.[24] In 1607 Captain John Smith established his reputation at Jamestown. Despite Jamestown's failure six years later, the British began settling the Atlantic coast and exploring inland, especially along large river tidewaters. Gradually, three major passages developed through these mountains that funneled settlers into the Greater Mississippi Valley. The northern route was the Mohawk Trail, extending from Fort Orange at present-day Albany up the Hudson River, then west by way of the Mohawk River to Lake Ontario. The central route was the Kittanning Trail, which utilized Susquehanna tributaries to reach the Conemaugh and Allegheny Rivers, which settlers then followed into the Ohio River.[25] The southern route was the famous Cumberland Gap, named in 1750 by young Thomas Jefferson's guardian, Thomas Walker, and located near the junction of Virginia, Kentucky, and Tennessee. Five years later, Daniel Boone carved the Wilderness Road through the Cumberland Gap, and from 1764 to 1774, British colonists settled this region at a rate of seventeen miles per year,[26] displacing resident Indian tribes such as the Lenape (Delaware) westward, across the Mississippi River.

The Mississippi is the Great Grand River of the United States. While at 2,320 miles it is the second longest river in the United States (curiously, its tributary, the Missouri, is 121 miles longer[27]), during the eighteenth and nineteenth centuries, it was the main center of exploration and commercial outflow in the heart of the country. St. Louis was founded in 1763 at the confluence of the Missouri and Mississippi Rivers by the French fur trader Pierre Laclède, who named the settlement for King Louis XI. The shallow, meandering Missouri became the highway to the West.

From its beginning, St. Louis held the key to the American Northwest—the area between Spanish- and French/English-occupied territories. This was an undefined land in an unstable time. In 1762 Louis XV secretly transferred control of the Louisiana Territory, including St. Louis, to his cousin, Charles III of Spain.[28] France then regained the region back in 1800, prior to selling it to the United States three years later as the Louisiana Purchase. The price—$11.25 million plus a cancelation of $3.75 million in debts—helped offset the costs of Napoleon's failed Caribbean ambitions. France's offer came at the perfect time, for Jefferson, alarmed about British intentions in the Pacific Northwest, needed to secure this region.

In the mid-1770s, Englishman Peter Pond had explored the Upper Mississippi River Basin in present-day Minnesota and Wisconsin. By 1780 the aggressive Pond had pushed beyond the range of the existing fur trade around the Great Lakes, into the Athabasca Basin of present-day Saskatchewan and Alberta "to explore a country hitherto unknown but from an Indian report."[29] Pond was convinced at his farthest point that he was but a week's travel from the Pacific Ocean, and in 1789 his colleague Alexander Mackenzie set out to prove it. On June 3, Mackenzie left Lake Athabasca and, traveling down the river that now bears his name, arrived at the Arctic Ocean five weeks later, on July 12. Realizing his mistake, Mackenzie corrected it. In 1792 he found a route to the Pacific Ocean through the Canadian Rockies and down the Fraser River. Mackenzie reached the Pacific on July 17, 1793, and thus became the first European to cross the North American continent north of Mexico. Mackenzie was honest, fair, and responsible. None of his men deserted or died during his explorations, nor did his parties kill any Native Americans.[30] In 1801 Mackenzie published his journals from the two expeditions, which Jefferson read. Within two years, the third United States president purchased the Louisiana Territory and ordered Lewis and Clark's expedition.

Jefferson's instructions to Lewis and Clark contained equal amounts of advanced knowledge of United States geography and great naïveté, as follows (the spelling is Jefferson's):

> . . . the North river or Rio Bravo which runs into the gulph of Mexico, and the North river, or Rio Colorado, which runs into the gulph of California, are understood to be the principal streams heading opposite to the waters of the Missouri, and running Southwardly. . . . The northern waters of the Missouri are less to be enquired after, because they have been ascertained to a considerable degree, and are still in a course of ascertainment by English traders & travellers. but if you can learn anything certain of the most northern source of the Missisipi & of it's position relative to the lake of the woods, it will be interesting to us. some account too of the path of the Canadian traders from the Missisipi, at the mouth of the Ouisconsin river, to where it strikes the Missouri and of the soils & rivers in its course, is desireable.[31]

A year later Lewis and Clark embarked from St. Louis on their Voyage of Discovery. Five biological/sociological features of Lewis and Clark's journals stand out. The first is the vegetational structure of the fire-dominated landscape of the Great Plains. The second is the collection of plants and animals at identified geographic locations—many of which no longer occur at these sites, or indeed across much of their historic range. The third is the large number of animals, especially the big mammals—bison, deer, and elk; from their accounts, it is easy to see why the Great Plains has been referred to as the North American Serengeti. The fourth is the attention to detail in their descriptions of animal specimens, especially new species. And the fifth is their observations of, and interactions with, an already depleted Native American population. Although Lewis and Clark's biological specimens were never used to their potential,[32] their tally of new species ran to 178 plants and 122 animals.[33] Of the animals collected, only one specimen, a Lewis's woodpecker (*Melanerpes lewis*), survives intact, at Harvard's Museum of Comparative Zoology, where it was transferred in 1914.[34]

The Philadelphia Academy of Natural Sciences, founded in 1812, was the first great American scientific institution. Early on, it was the chief American repository for new biological specimens being shipped east; it is where Lewis and Clark sent their plants and animals. Philadelphia was the political and intellectual center of American settlement during the days preceding the Revolution and for a half century following. Benjamin Franklin founded the American Philosophical Society there in 1743, and Jefferson was its president for four years, before he went to Washington to become the country's president in 1801, the year after the capital was moved from Philadelphia.[35]

Slowly, Lewis and Clark's species descriptions appeared in print. As early as 1808, their birds were being described and illustrated by Alexander Wilson. In 1814 the botanist Frederick Pursh described seventy-seven plant species in his *Flora Americae Septentrionalis*, and George Ord described ten animal species, including grizzly bears and pronghorn antelopes. Also in 1814 a condensed and filtered version of Lewis and Clark's journals was published. A planned second volume, covering the Voyage of Discovery's natural history and geological accomplishments, was never published because the editor, Benjamin Smith Barton—Lewis's mentor—was too ill.[36] As Jefferson anticipated, commercial enterprises followed the Voyage of Discovery. Two years after Lewis and Clark's return, the German-born John Jacob Astor founded the Western Division of the American Fur Company to break American dependence on the Hudson's Bay Company's monopoly on beaver pelts.[37]

In July 1806, before Lewis and Clark's return, General James Wilkinson, governor of the Louisiana Territory, ordered Zebulon Pike to explore the source of the Red River, the legal boundary between American and Spanish lands. Pike's outfit discovered the mountain named for its leader and pioneered the Santa Fe Trail. Teddy Roosevelt admired Pike,[38] but "nothing that Pike ever tried to do was easy, and most of his luck was bad."[39] By Pike's own admission, he was an untrained naturalist and lacked the "time and placidity of mind" required to study the plants and animals he encountered. During his expedition, Pike and some of his men were captured by the Spanish and held for several months in Chihuahua before being released.[40] Roosevelt's admiration notwithstanding, Pike's report, published in 1810, was considered "poorly organized, unreliable . . . scientifically and geographically incorrect, and in many places dishonest."[41]

These early government-sponsored expeditions have been described by historians as explorations, but in fact they were driven by commerce. Working out of St. Louis, Manuel Lisa and his partner Andrew Henry established the Missouri Fur Company in 1809, which sent expeditions up the Missouri River to open this region to trade. The Missouri Fur Company conducted their operations out of Fort Lisa, near present-day Omaha, and more distantly out of Fort Raymond on the Yellowstone River, in eastern Montana. Lisa's trappers combed the Rockies. For example, John Colter (a member of the Voyage of Discovery, once reprimanded for his behavior) discovered the geysers of Yellowstone during an 1807–8 trip. Lisa's men also found a route down the Rockies to Santa Fe, the Spanish center of commerce, during the 1811 exploration led by Jean Baptiste Champlain. The Missouri Fur Company's explorations eventually accumulated so much information useful to William Clark's mapping project of the West that in 1814 the U.S. government rewarded Lisa by appointing him as agent for Indian affairs in the upper Missouri region.[42]

Lisa's Missouri Fur Company both assisted and enabled exploration, but the major players in the trade in beaver pelts were the American Fur Company, founded in 1808 by Astor, and the Rocky Mountain Fur Company, formed in 1822.[43] Early in its history, the American Fur Company used the Missouri and Columbia Rivers and their tributaries to transport supplies and equipment in and furs out—they were, in essence, voyageurs (boatmen). In 1811 the American Fur Company sent a party led by Wilson Price Hunt up the Missouri. They eventually made their way, following the trail of Lewis and Clark, to the Pacific, where they founded the town of Astoria. Traveling with Hunt as far as the Arikara villages in present-day North Dakota[44] were the British botanists Thomas Nuttall and John Bradbury. In 1817 Bradbury wrote an account of this trip in *Travels in the Interior of America in the Years 1809, 1810, and 1811.*[45] In addition to describing plants collected by Lewis and Clark, Pursh's *Flora Americae Septentrionalis* (1814) described, without permission, plants collected by Nuttall and Bradbury.[46] The American Fur Company improved traditional methods of river transportation by introducing steamboats to the upper Missouri. Astor's company operated fixed trading posts and relied on Indian trappers; it traded in all furs, as well as feathers, meat, and lard.[47] Largely because of its tendency to trap out areas and move on, by 1833 beaver had become scarce, and bison robes became their primary commodity.[48]

In contrast, the Rocky Mountain Fur Company worked on land, away from rivers, using pack trains up the valley of the Platte River and through the South Pass.[49] (In 1812 American Fur Company trapper Robert Stuart, traveling back east from Astoria, found a way through the Rockies at South Pass, Wyoming, in what was to become the route of the Rocky Mountain Fur Company and, a few decades later, the "Oregon Trail," and then a few decades after that the route of the Union Pacific Railroad.) The Rocky Mountain Fur Company maintained no fixed posts, traveled in small parties, caught their own beaver, and traded with Native Americans as chance opportunities arose or at their annual rendezvous, which occurred during the summer at a predetermined time and place. Their employees became the prototypical and highly celebrated mountain men, including Jim Bridger, Kit Carson, and Tom Fitzpatrick. They would opportunistically trap marten, otter, mink, and fox, but their interest in bison did not extend much beyond personal use. Gradually, the distinctions between the American and Rocky Mountain Fur companies dissolved and by 1833 had disappeared. By then, through their complimentary approaches, they had opened the West.

The federal government sponsored other explorations concurrent with, but less well known than, Lewis and Clark's. In 1804 William Dunbar led a three-month expedition with George Hunter as naturalist, up the Ouachita River, a tributary of the lower Red River. There, they found the Hot Springs area of central Arkansas. Dunbar was an inventor, and he worked out a novel way of calculating longitudes based on lunar altitudes; he also designed flat-bottomed boats for exploring shallow prairie rivers such as the Missouri. In April 1806 Thomas Freeman led a second Red River expedition. Peter Custis was the naturalist, and together they traveled upriver 615 miles before being forced by Native Americans to turn back. These two Red River surveys were the first to deploy civilian scientists (Hunter and Custis). Despite not achieving their goal of clarifying the boundary between the Louisiana Purchase and Spanish-held territories, they did assemble accurate, detailed maps, as well as information on the geography, biology, and Native Americans of the region.[50]

Wallace Stegner noted what ship-bound British naturalists had already experienced,[51] that hazardous and unsecure conditions associated with early expeditions (e.g., traveling by boat over water—or, worse, on fast-flowing rivers—or by horseback and wagon; having only tenuous shelter in the evening and during storms; and coping with subfreezing conditions) caused biological specimens to wet and rot or be lost. As John Frémont related in a letter to the botanist John Torrey:

> Though in the course of our journey the Bales of plants had been twice wet, yet they were in very beautiful order when we encamped on the upper waters of the Kansas on the 13th of July, in the course of which night it began to rain violently & towards morning the river which was over 100 yards wide suddenly broke over its banks, becoming in less than 5 minutes more than half a mile in breadth. Everything we had was soaked. We were obliged to move camp to the Bluffs in a heavy rain which continued for several days and one fine collection was entirely ruined. . . . I brought them along and such as they are I send them to you. They are broken up & mouldy and decayed, and today I tried to change some of them, but found it better to let them alone.[52]

This sort of mischance happened to every collector. Prince Maximilian of Wied-Neuwied lost most of his plant and animal specimens when the steamboat *Assiniboine* caught fire and burned on the Missouri River; Pike's field notes were confiscated by Spaniards; a portion of Thomas Say's and Edwin James's field notes were carried off by deserters; Joseph Nicollet's botanist, Charles A. Geyer, lost his plant collection in shipment from Fort Snelling to St. Louis;[53] and Charles Parry's specimens from the Mexican Boundary Survey were destroyed in a warehouse fire in Panama.[54] Some threats recurred: Nuttall had trouble keeping his lizard and snake specimens hydrated because Native Americans kept drinking the alcohol from his collecting jars.[55]

Aware of these dangers, Jefferson recommended to Lewis and Clark that they write field notes on birch bark, "it being little liable to injury from wet and other accidents."[56]

A decade and a half after Lewis and Clark's return, U.S. Army engineer Stephen H. Long led a government-sponsored expedition up the Missouri River, this time from St. Louis to the Platte River then west to the Colorado Front Range.[57] Their major objectives were to explore the region between the Mississippi River and the Rocky Mountains, the Missouri and its tributaries, and the Red, Arkansas, and Mississippi Rivers.[58] Long had with him the botanist Edwin James and the zoologist Thomas Say, who, as a result of this expedition, were the first to provide scientific descriptions and formal Latinized names of plants and animals of the American High Plains.[59] Also with them was assistant naturalist and artist Titian Peale and landscape artist Samuel Seymour.[60] The Long Expedition was the first western survey to include trained naturalists. While these explorers carried unpublished maps made by Clark and Pike, as well as the journals of Lewis and Clark,[61] their report, released in

1823, was published prior to the delayed full report of the Voyage of Discovery, making Long's discoveries better known at the time.[62] Further, the Long Expedition report was illustrated, and for the first time curious easterners could see images of the Great Plains and the Rockies, along with portrayals of native North Americans.[63]

The Long Expedition of 1819–20 attracted attention for another reason—their characterization of the land between the Mississippi River and the Rockies as the "Great American Desert."[64] Their assessment was based on observations such as those of their botanist, James, who was confident, if not exactly accurate, in his opinions:

> The traveller journeys, for weeks in succession, over a dreary and monotonous plain, sparingly skirted and striped with narrow undulating lines of timber, which grow only along the margins of considerable streams of water. In these boundless oceans of grass, his sensations are not unlike those of the mariner, who beholds around him only the expanse of the sky and the waste of waters.[65]

James echoed and may have borrowed from Coronado, who described the Llano Estacado as "plains so vast, that I did not find their limit anywhere I went . . . with no more land marks than if we had been swallowed up by the sea."[66] (James Fenimore Cooper in *The Prairie* and Willa Cather in *My Antonia* make the same analogy. Cooper wrote: "So very striking was the resemblance between the water and the land, that, however much the geologist might want to sneer at so simple a theory, it would have been difficult for a poet not to have felt that the formation of the one had been produced by the subsiding dominion of the other."[67] Cather offered, succinctly, but with more depth: "I felt the grass was the country, as the water is the sea."[68]) James then went to the opposite hydric extreme and called the western plains: "arid and sterile," "barren as the deserts of Arabia," and "a desolate and disgusting tract of country," one that "must remain forever desolate."[69]

James's assessment—shared by Long and read by many in the East—slowed immigration into the region until after the Homestead Act of 1862 was passed and transcontinental railroads were built, and for this, Long and his expedition have been widely criticized.[70] While historians still bicker over whether Long was correct about the idea of the Great Plains as desert,[71] western naturalists long ago settled the issue. In the last quarter of the nineteenth century, John Wesley Powell recognized the approximate boundary between the arid lands (fewer than twenty inches of annual rainfall) in the rain shadow of the Rocky Mountains and the arable, wetter lands to the east as the 100th meridian. The mid-twentieth-century prairie ecologist John Weaver considered the region east of the 100th meridian to be "Prairie," the region west to be "Great

Plains."[72] Today, most scientists consider the Great Plains to include all grass-lands east of the Rockies and west of a longitude corresponding roughly to the western borders of Minnesota and Iowa running south through Omaha and Kansas City, into east Texas.

In 1823 Long led a second expedition to the upper Midwest. His party first traveled up the Minnesota River, then down the Red River north into Canada, which belatedly satisfied Jefferson's desire to better know the waters of the Upper Mississippi River Basin. It would fall to Joseph Nicollet two decades later to complete the task.

In 1832 American author and diplomat Washington Irving was seeking to reestablish his credentials in the eyes of a skeptical American public after spending seventeen years in Europe. He offered the universal "we" to his specific "me" issue: "We send our youth abroad to grow luxurious and effeminate in Europe; it appears to me that a previous tour on the prairies would be more likely to produce that manliness, simplicity, and self-dependence most in unison with our political institutions."[73] The solution for Irving was to join a government mission to monitor the settling of displaced Indian tribes being pushed west by encroaching Euro-American settlement. Beginning that October, Irving spent a month in what is now Oklahoma and wrote this about the early nineteenth-century expansion of settlers:

> It is surprising in what countless swarms the [honey] bees have overspread the Far West, within but a moderate number of years. The Indians consider them the harbinger of the white man, as the buffalo is of the red man; and say that, in proportion as the bee advances, the Indian and buffalo retire. We are always accustomed to associate the hum of the bee-hive with the farmhouse and flower-garden, and to consider those industrious little animals as connected with the busy haunts of man, and I am told that the wild bee is seldom to be met with at any great distance from the frontier. They have been the heralds of civilization, steadfastly preceding it as it advanced from the Atlantic borders, and some of the ancient settlers of the West pretend to give the very year when the honey-bee first crossed the Mississippi. The Indians with surprise found with ambrosial sweets, and nothing, I am told, can exceed the greedy relish with which they banquet for the first time upon this unbrought luxury of the wilderness.[74]

Four years later, in 1836, Irving based chapters XIII–XXXVIII of his book *Astoria* on the 1811 Pacific Fur Company expedition up the Missouri, in which Nuttall and Bradbury served as botanists as far as the Arikara (Bradbury) and Mandan (Nuttall) villages located in present-day southwestern North Dakota.

Prior to the construction of the transcontinental railroads, the main route to the Plains was on commercial vessels plying the Missouri River. In 1832, the year of Irving's manliness trek, the American Fur Company's steamboat *Yellowstone* made the first trip to Fort Union at the confluence of the Yellow-stone and Missouri Rivers. On March 24, 1833, Prince Maximilian of Wied-Neuwied arrived at St. Louis with his servant Dreidoppel and the painter Karl Bodmer. Bernard DeVoto described Maximilian as a thin, worn man of fifty, excitable, choleric, with a gift of invective and an unleavened Prussian accent. He had fought against Napoleon and been captured. He had not wanted to be a soldier—he wanted to be a scientist. After the war, he spent two years exploring Brazil, studying the natives and its natural history. His book *Reise nach Brasilien*,[75] published in 1820, was well received, and he came to North America to engage in a similar study. Although Maximilian had experience in botany, zoology, geology, paleontology, and meteorology, his primary in-terest was ethnology. He traveled using the name Baron von Braunsberg, and by the time Maximilian reached North America, the hardships he had experi-enced while serving in the military and on his tropical geographical explora-tions had cost him his teeth.[76]

In St. Louis, Maximilian marveled at the Africans, saw his first western Na-tive Americans, and soaked up the strangeness of a cosmopolitan river town in the center of an empty continent. Before he departed, he met Irving and sat while George Catlin painted his portrait. Then, on April 10, at 10:30 AM, ac-cording to his journal,

> . . . the steam engine on the Yellow Stone was started after the group had as-sembled on board. Mr Chouteau, his daughters, and other ladies accompanied us on the journey today. The Yellow Stone's flags were fluttering, the American one aft and a narrower one . . . with the initials of the American Fur Company. Several cannon shots had been fired, whereupon the residents of St. Louis as-sembled en masse.[77]

Catlin, the famous painter of and advocate for Native Americans, also used American Fur Company steamers to get around. By the time Catlin reached the Mandans, located in present-day North Dakota, in 1833, these friendly Native Americans had been interacting—in every sense of the word—with whites for over a century. Their fair skin and occasional blue eyes—fur-trader genes—led him to believe these were the "Welsh Indians" of frontier legend.[78] When, in 1835, the French-born cartographer Nicollet viewed Catlin's work after returning to St. Louis, he thought it pleasurable and interesting.[79]

Nicollet was social and made friends quickly. In the mid-1830s, he met George W. Featherstonhaugh, who in 1835 conducted a geological reconnais-sance from Green Bay across Wisconsin, and then up the Minnesota River to the Prairie Coteau in South Dakota. Featherstonhaugh greatly influenced the

French surveyor, emphasizing the importance of establishing accurate car-
tographic coordinates.[80] From 1836 to 1837, Nicollet traveled up the St. Croix
River to Lake Superior, then west to the Red River and up it (south) to the
Minnesota River, then down the Minnesota and home to Fort Snelling.

Nicollet's map of the Upper Mississippi caught the attention of John J.
Abert, head of the Topographical Bureau of the War Department, in Washington, DC. On February 6, 1838, Nicollet wrote to his friend and former host,
the American Fur Company's agent Henry H. Sibley: [81] ". . . your government
desires to retain me to explore the beautiful regions west of the Mississippi,
and between the St. Peter's and the Des Moines [Rivers]." Portions of Nicollet's proposed route had been traveled by Long (in reverse) in 1823 and by
Featherstonhaugh in 1835, but their maps of this region lacked the accuracy
of Nicollet's maps. In addition to cartography, Nicollet's responsibilities included "recording data on climate, soils, geology, botany and zoology,"[82] as
well as reporting on the number and attitude of the remaining Native Americans in this region, which was known to have agricultural potential and was
attracting European settlers.

Nicollet was the first to lead an exploring party under the auspices of the
newly formed Corps of Topographical Engineers. He employed the most modern cartographic techniques, including not only accurate sightings of latitude
and longitude, but also a barometric measurement of elevation. As Nicollet
detailed:

> The reconnaissance of the country traversed each day, or rather the survey
> of our route, by land or by water, was made by making the magnetic bearing
> of every point, by estimating its distance, and by making, as we went, a con-
> nected sketch or bird's-eye view of the whole, and very often including distant
> points of importance, indicated to us by the guides, to which one of us always
> went to take note of.[83]

Nicollet would make the finest early maps of the area between the Missouri
and Mississippi Rivers and set the standard for the subsequent cartography
of the West. As Norman Maclean observed, "If you don't know the ground,
you're probably wrong about nearly everything else."[84] Nicollet was rarely
wrong about the ground.

Nicollet's personality made him popular among the French in St. Louis.
There, he met the British explorer and bison hunter[85] Sir William Drummond Stewart.[86] Captain Stewart, like Maximilian a veteran of the Napole-
onic Wars, was quite the dandy in his white jacket and Panama hat, but he
was also a tough guy—an expert rider and shooter. According to the historian
Tom McHugh, Stewart, along with his guides, left the prairie bordering the

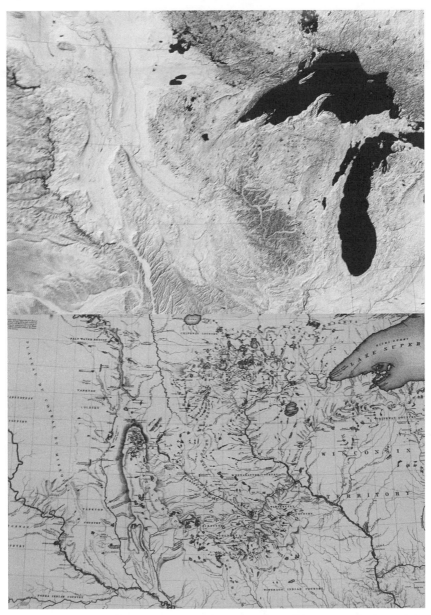

FIGURE 3A & 3B. A comparison of the Upper Midwest using Lidar satellite mapping technology with Joseph Nicollet's ground-based mapping results. By 1830 standards, Nicollet got it right. (a) From the David Rumsey Map Collection (www.davidrumsey.com). (b) Created by Dr. Alisa Gallant, USGS.

Platte River "strewn for miles" with carcasses of slain bison.[87] DeVoto described Stewart as five feet nine, with "black hair," a "luxuriant mustache," and a beaked nose. He was "a man with strong prejudices and equally strong appetites, which he had freely indulged." DeVoto suggested that his face was "thin from the free use of New Orleans brandy."[88]

Stewart deeply admired the lifestyle of the American frontier, although he might have been a British spy.[89] After spending six straight summers in the West, he came back once more, in 1843. When Stewart returned to Scotland after this trip, and for good, he became a novelist—in DeVoto's opinion, "an unbelievably bad one."[90]

The Settlers

As fur traders pushed west, pioneers followed and settled the Middle West. At the time of Lewis and Clark's Voyage of Discovery, the East Coast—New York north along the Hudson and up the Mohawk Valley to Schenectady, eastern and southern Pennsylvania, Virginia across the Shenandoah Valley, as well as the Carolinas and eastern Georgia—was largely British.[91] Beyond this center, small populated areas had sprung up in Kentucky, Tennessee, and Ohio.[92] The Midwest had a population of about 100,000, one-fortieth of the United States' total.[93]

After the War of 1812, steamboats made river travel easier and more reliable, and this convenience showed in census data.[94] Frederick Jackson Turner described two waves of early eighteenth-century midwestern immigration. The first comprised Scottish-Irish descendants from the interior of the South—primarily from Kentucky, Virginia, and North Carolina—who "with ax and rifle in hand had cut their clearings in the forest, raised their log cabins, fought the Indians and by 1830 had pushed their way to the very edge of the prairies along the Ohio and Missouri Valleys, leaving unoccupied most of the Basin of the Great Lakes." Abraham Lincoln's father, Thomas, left Kentucky for Indiana in 1816.[95]

These early settlements were tied to rivers. The Midwest had a flow-through economy. Pittsburgh, by way of the Ohio River, furnished most of the supplies, and the Mississippi River through New Orleans provided the principal outlet for goods, including crops. The major cities of the Old National, or Cumberland, Road (in the old days Highway 40; now, roughly, Interstate 70)—Columbus, Indianapolis, and Vandalia connecting to St. Louis—marked the northern border of the early southern stream of midwestern immigration.[96]

The opening of the Erie Canal, in 1825, unlocked the northern migration. Settlers populated the southern tiers of counties across present-day Michigan and Wisconsin, the northern counties of Illinois, parts of the northern and central areas of Indiana, and southeastern Iowa. Again Turner: "oak openings and prairies gave birth to the cities of Chicago, Milwaukee, St. Paul, and Minneapolis, as well as to a multitude of lesser cities."[97] By 1840 the Midwest had a population of over 6 million, more than one-third of the population of the United States.[98] Censuses showed an uneven advance of frontier, with fingers of settlement punctuating gaps of wilderness. There was, however, a pattern to the process of midwestern settlement:

> The buffalo trail became the Indian trail, and this became the trader's "trace"; the trails widened into roads, and the roads into turnpikes, and these in turn were transformed into railroads. . . . The trading posts reached by these trails were on the sites of Indian villages which had been placed in positions suggested by nature; and these trading posts, situated so as to command the water systems of the country, have grown into such cities as Albany, Pittsburgh, Detroit, Chicago, St. Louis, Council Bluffs, and Kansas City.[99]

And there was a pattern to the types of people who arrived:

> First comes the pioneer. . . . He builds his cabin, gathers around him a few other families of similar tastes and habits, and occupies till . . . the neighbors crowd around, roads, bridges, and fields annoy him, and he lacks elbow room. The preëmption law enables him to dispose of his cabin and cornfield to the next class of emigrants; and, to employ his own figures, he "breaks for the high timber," "clears out for the New Purchase." . . . Another wave rolls on. The men of capital and enterprise come. . . . The small village rises to a spacious town or city; substantial edifices of brick, extensive fields, orchards, gardens, colleges, and churches are seen. . . . Year by year the farmers . . . lived on soil whose returns were diminished by unrotated crops. . . . Their growing families demanded more lands, and these were dear. The . . . easily tilled prairie lands compelled the farmer either to go west and continue the exhaustion of the soil on a new frontier. . . . Thus the demand for land and the love of wilderness freedom drew the frontier ever onward.[100]

These people and processes gave rise to the homegrown field biologists we are about to consider. By 1850 about one-sixth of midwesterners were of New England/New York birth, about one-eighth of southern birth, and a similar fraction of foreign birth. Germans were more numerous than Scandinavians, who were more numerous than Scots. Almost three-fifths of midwesterners were natives of the Midwest, and over one-third of midwesterners lived in Ohio. Nebraska and Kansas marked the frontier. Minnesota and Wisconsin still exhibited frontier conditions, although the Treaty of Traverse des Sioux, in 1851, opened over 20 million acres of arable land, increasing Minnesota's population 27-fold in the decade from 1850 to 1860.[101]

After 1870 transcontinental railroads, the Homestead Act, and the realization that Stephen Long's "Great American Desert" could be farmed and ranched attracted an increasing tide of immigrants to the Far West[102]:

> The Northern Pacific and the Great Northern Railway thrust out laterals into these Minnesota and Dakota wheat areas from which to draw the nourishment for their daring passage to the Pacific. . . . The railroads sent their agents and their literature everywhere, "booming" the "Golden West"; the opportunity for economic and political fortunes in such rapidly growing communities attracted multitudes of Americans whom the cheap land alone would not have tempted. . . . Nebraska's population was 28,000 in 1860 . . . 1,059,000 in 1890.[103]

There was a contrary view of the motivation of pioneers, born out of the discomfort of the conservative eastern establishment. According to Turner, Timothy Dwight IV of Yale wrote in his 1823 book *Travels in New England and New York*:

> The class of pioneers cannot live in regular society. They are too idle, too talkative, too passionate, too prodigal, and too shiftless to acquire either property or character. They are impatient of the restraints of law, religion, and morality, and grumble about the taxes by which the Rulers, Ministers, and Schoolmasters are supported. . . . [T]hey become at length discouraged, and under the pressure of poverty, the fear of the gaol, and consciousness of public contempt, leave their native places and betake themselves to the wilderness.[104]

As Turner points out, it was among these "plain people," as Abraham Lincoln called them, that the sixteenth president grew to manhood. Emerson once said of Lincoln, "He is the true history of the American people in his time." By the beginning of the Civil War, midwesterners were a dominant national force. Lincoln was in office, Grant and Sherman were the Union's best generals, and the region provided over one-third of the Union troops. After Lincoln was assassinated, four of the next six presidents (Ulysses S. Grant, Rutherford B. Hayes, James A. Garfield, and Benjamin Harrison) came out of the Midwest.[105]

The Homestead Act accelerated settlement. Signed into law by Lincoln on May 20, 1862, it provided 160 acres of land for anyone twenty-one years old or older, or who was the head of a family, who could prove residency, and who had never taken up arms against the U.S. government. These provisions were extended to freed slaves and women.

Lincoln signed a second critical piece of legislation six weeks later, on July 2. The Morrill Act provided "liberal and practical education" in the "agriculture

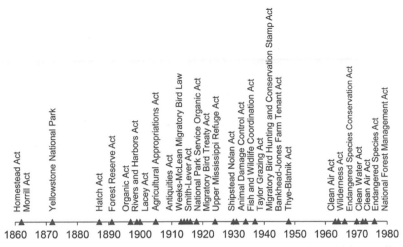

FIGURE 4. The chronology of key federal legislation impacting settlement, the establishment of land-grant universities, land set-asides, and environmental protection.

and the mechanic arts" to promote the industrial classes in the several pursuits and professions in life. Under this act, each qualifying state received a total of 30,000 acres (120 km²) of federal land, either within or contiguous to its boundaries, for each member of Congress representing the state according to the 1860 census. This land, or the proceeds from its sale, was to be put toward establishing and funding the educational institutions described above. If the state deemed the federal land within its boundaries was insufficient to meet its land grant,[106] that state was issued "scrip" that allowed it to select federal lands from elsewhere to fund its institution. The state of New York used this clause to select western federal old-growth timber holdings, which it then sold off to bankroll Cornell. After the Civil War, the provisions of the Morrill Act were offered to the former Confederate states, as well as to every state and territory founded after 1862.

On September 11, 1862, Iowa was the first state to accept the terms of the Morrill Act, which provided needed funding for the State Agricultural College, in Ames. Land-grant universities became a characteristic of the Midwest, endorsed and supported by the community as a whole, rather than by prominent benefactors. Henry S. Pritchett, president of the Carnegie Foundation, called the state university "a conception of education from the standpoint of the whole people. . . . If our American democracy were to day called to give proof of its constructive ability, the State University and the public school system which it crowns would be the strongest evidence of its fitness which it could offer."[107] A provision in the Constitution of Indiana, written in

1816, provided for a "general system of education ascending in regular grada-
tions from township schools to a State University, wherein tuition shall be
gratis and equally open to all."[108]

The Thinkers

In the mid-1830s, at about the same time that the trade in beaver pelts was
collapsing, a Unitarian minister turned lecturer named Ralph Waldo Emerson
published an essay called *Nature*.[109] Emerson was a prodigy. He entered
Harvard at age fourteen and was preaching at twenty-three. After his young
wife died, Emerson had a crisis of faith and took time off to travel across
Europe.[110] There, he fell under the influence of the Scottish historian and
philosopher Thomas Carlyle, as well as Samuel Taylor Coleridge and William
Wordsworth of Britain's Romantic movement. After Emerson returned, he
remarried and settled in Concord, Massachusetts. A year later he began meeting with like-minded people such as Bronson Alcott, Orestes Brownson, and
Margaret Fuller, and together they began the Transcendentalist movement.

Transcendentalists made up the United States' first coherent intellectual
movement. From the early 1830s until the early 1850s, adherents met in one
another's homes, attended one another's lectures and sermons, and read and
reviewed one another's writings.[111] As George Ripley described in 1840, Transcendentalists "believe in an order of truths which transcends the sphere of
the external senses. Their leading idea is the supremacy of mind over matter."
Ripley continued, "There is a light, they believe, which enlighteneth every
man that cometh into the world; there is a faculty in all, the most degraded,
the most ignorant, the most obscure, to perceive spiritual truth."[112] Transcendentalists advocated for educational reform, women's rights, labor, temper-
ance, prisoners' rights, and the abolition of slavery. Henry David Thoreau was
Emerson's protégé; Philip Gura called Thoreau's *Walden* the "literary masterpiece of the Transcendentalist movement."[113]

Following the Civil War, Emerson was idolized as America's philosopher
for his belief in self-reliance—fostering the notion of the rugged individual.[114]
But it was Emerson's belief that "Nature mirrors the divine"[115]—taken up
by Thoreau along the East Coast and Muir, a generation later, on the West
Coast—that helped to kindle what would become the preservationist arm
of the environmental movement. Emerson laid the foundation for a holistic
view of nature:

> When we speak of nature . . . [we] mean the integrity of impression made by
> manifold natural objects. It is this which distinguishes the stick of timber of

the wood-cutter, from the tree of the poet. The charming landscape which I saw this morning, is indubitably made up of some twenty or thirty farms. Miller owns this field, Locke that, and Manning the woodland beyond. But none of them owns the landscape. There is a property in the horizon which no man has but he who can integrate all the parts, that is, the poet. This is the best part of these men's farms, yet to this their land-deeds give no title.[116]

Thoreau's approach to nature is found in a passage that may have inspired one of Edward Abbey's most famous quotes (see p. xi):

Live in each season as it passes; breathe the air, drink the drink, taste the fruit, and resign yourself to the influences of each. Let them be your only diet drink and botanical medicines. In August live on berries, not dried meats and pemmican, as if you were on shipboard making your way through a waste ocean, or in a northern desert. Be blown by all the winds. Open all your pores and bathe in all the tides of Nature, in all her streams and oceans, at all seasons. Miasma and infection are from within, not without.[117]

As with Emerson, Thoreau demonstrated an excellent eye for natural phenomena:

As I come over the hill, I hear the wood thrush singing his evening lay. This is the only bird whose note affects me like music, affects the flow and tenor of my thought, my fancy and imagination. It lifts and exhilarates me. It is inspiring. It is a medicative draught to my soul. It is an elixir to my eyes and a fountain of youth to all my senses. It changes all hours to an external morning. It banishes all trivialness. It reinstates me in my dominion, makes me the lord of creation, is chief musician of my court. . . . This thrush's song is a *ranz des vaches* to me. I long for wildness, a nature which I cannot put my foot through, woods where the wood thrush forever sings, where the hours are early morning ones, and there is dew on the grass, and the day is forever unproved, where I might have a fertile unknown for a soil about me.[118]

Thoreau summarized Transcendentalist thinking on conservation issues, when he offered:

Every larger tree which I knew and admired is being gradually culled out and carried to mill. I see one or two more large oaks in E. Hubbard's wood lying high on stumps, waiting for snow to be removed. I miss them as surely and with the same feeling that I do the old inhabitants out of the village street.[119]

The emphasis on individualism that propelled Emerson to become antebellum America's philosopher laureate permeates this Transcendentalist view of nature. While intellectuals such as Emerson and Thoreau, and sensitive geniuses like John Muir, had no problem providing vivid descriptions of the worlds they perceived, there is variation among humans in both the expression and the reception of this world. Many people understand and share the joy that Thoreau evokes in his journal entries and in *Walden*. But if one happens to be distant from Concord, as most are, to experience this same joy firsthand, they might want to know exactly which plant, frog, bird, and mammal species Thoreau was contemplating. In many cases, it is impossible from his descriptions to know. In contrast, science—a group effort—demands communication in a standardized fashion; it needs a common language.

The nature and norms of debate may have been different in the mid-1800s, but by today's standards, it is unfortunate that both Emerson and Thoreau sought to promote their Transcendentalist views by diminishing the perspectives provided by science—a facet of early Transcendentalism that has been largely ignored or dismissed.

On March 5, 1853, Thoreau wrote:

The Secretary of the Association for the Advancement of Science [Thoreau was never a member] requests me, as he probably has thousands of others, by a printed circular letter from Washington the other day, to fill the blank against certain questions, among the most important one was what branch of science I was specially interested in, using the term science in the most comprehensive sense possible. Now, though I could state to a select few that department of human inquiry which engages me, and should be rejoiced at an opportunity to do so, I felt that it would be to make myself the laughing-stock of the scientific community to describe or attempt to describe to them that branch of science which specially interests me, inasmuch as they do not believe in a science which deals with the higher law. So I was obliged to speak to their condition and describe to them that poor part of me which alone they can understand. The fact that I am a mystic, a transcendentalist, and a natural philosopher to boot. Now I think of it, I should have told them at once that I was a transcendentalist. That would have been the shortest way of telling them that they would not understand my explanations.

How absurd that, though I probably stand as near to nature as any of them, and am by constitution as good an observer as most, yet a true account of my relation to nature should excite their ridicule only! If it had been the secretary of an association of which Plato or Aristotle was president, I should not have hesitated to describe my studies at once and particularly.[120]

And Emerson declared:

The motive of science was the extension of man, on all sides, into Nature, till his hands should touch the stars, his eyes see through the earth; his ears understand the language of beast and bird, and the sense of the wind; and, through his sympathy, heaven and earth should talk with him. But that is not our science. These geologies, chemistries, astronomies, seem to make wise, but they leave us where they found us. The invention is of use to the inventor, of questionable help to any other, of no value to any but the owner. Science in England, in America, is jealous of theory, hates the names of love and moral purpose. There's a revenge for this inhumanity. What manner of man does science make? The boy is not attracted. He says, I do not wish to be such a man as my professor is. The collector has dried all the plants in his herbal, but he has lost weight and humor. He has got all snakes and lizards in his phials, but science has done for him also, and has put man into a bottle.[121]

Thoreau threw a last dart:

I suspect that the child plucks its first flower with an insight into its beauty and significance which the subsequent botanist never retains.[122]

Max Oelschlaeger gives the Transcendentalists a pass here, offering: "Thoreau is not advocating scientific book burning but is seeking a kind of cognitive balance, an 'Indian wisdom' that restores organic qualities to a world of scientific quantities and reintegrates human consciousness with the cognizable world."[123] Oelschlaeger missed observing this same tendency in Emerson. E. O. Wilson was aware of this tension and wrote in defense of the nineteenth-century science being done when the contrarian Transcendentalists were in ascension:

Biology could not have advanced without the collections of museums. . . . Absent their priceless resources, there would be no coherent system of classification, no way to identify the vast majority of organisms, no theory of evolution, no foundation for ecology.[124] Scientists would be traveling through a chaos of kaleidoscopic diversity. Biology would be a slender thread of experimental research reaching up from physics and chemistry and confined to a handful of organisms, probably known variously as "the fly," "the intestinal bacterium," "the cat," "the dog," "the human," "the corn," "the lily," and a few others.[125]

Perhaps if the Transcendentalists had approached their view toward life in a transcendent manner, they would have realized their beliefs were based in *individual* knowledge and understanding—the need to enrich the soul—as opposed to science, which is based in *group* knowledge and understanding— the need to communicate the facts of the experience (Oelschlaeger also missed this point).[126] A fully transcendent view might have appreciated that neither fundamental Transcendentalist nor classic scientific views by themselves were complete—that neither offered the inclusive knowledge and understanding necessary to satisfy *and communicate* the range of perspectives encompassed by the human mind hungry for a deep comprehension of the world.

Given the influence of the Transcendentalists in mid-nineteenth-century America, their stated hostility toward field biology and museum science (there are many more examples I could have chosen), and the exposure to Transcendentalism by East Coast students in their advanced intellectual climate, it comes as no surprise to learn than major East Coast museums and biological supply houses became populated with biologists from the less formally educated Midwest, where Transcendentalism held little sway. From this perspective, we can turn a Thoreau phrase back on him: "All this is perfectly distinct to an observant eye, and yet could easily pass unnoticed by most."[127]

The Institutions

As pioneers moved west, the requirement for explorations and expeditions in advance of this front gave way to the necessity of understanding the world surrounding their new settlements. This was an era of rapid advances in technology and transportation. It was also a time of emphasis on education and—through land-grant institutions—the means to disseminate it. At Harvard, Asa Gray, America's first academic botanist, witnessed most of this transformation.

Asa Gray was born November 18, 1810, in Sauquoit, New York.[128] After attending Fairfield Academy, he received his medical degree in 1830 and practiced medicine for a few years before immersing himself in his avocation of plant collecting. During the summer of 1830, while visiting New York, Gray left plants for John Torrey to examine. This gesture began a friendship and a series of collaborations on the botany of North America that would last until Torrey's death in 1873.

In 1826 Torrey had introduced the Natural System of plant classification to the United States,[129] and in August 1833 Torrey hired Gray to assist with his *Flora of North America* project. While working with Torrey, Gray published *Elements of Botany*. In part because of the recognition he received from this work, Gray was offered the first permanent paid professorship at the University of Michigan, which also happened to be America's first academic position dedicated to botany. Before Gray left New York for Europe to purchase books for the University of Michigan's library, he and Torrey studied plants collected by Thomas Nuttall during his three western sojourns.[130] As DeVoto described:

> Thomas Nuttall was no greenhorn but an innovator and pioneer with a distinguished reputation. . . . As a young man, he had accompanied the Astorians as far up the Missouri as the Arikara villages, somewhat south of the Mandan

villages. He and his fellow botanist, John Bradbury, who was also on that expedition, were the first trained scientists who ever entered the American Far
West—that was in 1811. The wonderland of unstudied plants had kept Nuttall
in an intoxication just short of frenzy. The voyageurs of Hunt's party labored
with him in vain, trying to tell him he was risking not only his life but also
Mr. Astor's money when he wandered off through prairies, thickets, and bottom land. But no Indians existed for Nuttall, there were no mischances, no
interruptions or delays, no possibility of disaster or death, there was only
the flora that no one had seen. So the voyageurs cursed him, decided he was
touched, and kept an eye on him when possible.[131]

Nuttall's second exploration, from 1818 to 1820, took him first to St. Louis,
down the Mississippi, and then up the Arkansas River, which at the time
marked the southern border between the United States and Spanish-held
lands. After returning, Nuttall spent two years in Philadelphia before moving to Cambridge to work up his collections and teach botany at Harvard.[132]
In 1834 he accompanied Nathaniel Wyeth's expedition up the Missouri. This
time he went all the way to the Oregon country. He wintered in the Sandwich
(Hawai'ian) Islands, returned to southern California, and headed east. Once
again, Nuttall took a position at the Philadelphia Academy until, strapped for
money and disillusioned with the attitude of Americans toward science, he
returned to England and assumed the casual lifestyle of the landed gentry.
Unlike Captain Stewart, Nuttall did not become a novelist.

In viewing Nuttall's specimens with Torrey in 1838, Gray was learning
from the master. Torrey had described Edwin James's plants from the 1819–
20 Long Expedition[133] and was the United States' expert on western botany.
However, most of these early specimens, especially the plants collected by
Nuttall and Bradbury during the Astorian Expedition, had been sent to
Sir William Hooker in Scotland.

When the University of Michigan sent Gray to Europe, he welcomed the
opportunity to visit the American plant collections. Gray's first objective
was to examine the European material and establish *Flora*—his collaboration with Torrey—as the definitive authority on North American plants; his
second was to develop collaborations with Old World colleagues and gain
their respect.[134] Gray met these goals by visiting Sir William Hooker and his
son Joseph in Glasgow, where he reviewed not only Nuttall's plants but also
those of David Douglas and others who had collected in the Pacific Northwest for the Hudson's Bay Company.[135] While in London, Gray gained access to Thomas Walter's eighteenth-century collection of Carolina plants and
John Clayton's Virginia plants. In addition, he examined Frederick Pursh's
collection, which formed the basis of his *Flora Americae Septentrionalis*.[136]

Gray also surveyed the herbarium of Carl von Linné (Carolus Linnaeus). In Paris, he examined the collections made by the French explorer André Michaux, who had journeyed through the southern United States between 1787 and 1807. Gray scrutinized the Paris material so thoroughly, he discovered a plant in an undescribed genus collected by Michaux in the "High Mountains of Carolina" in 1787. (In 1842 Gray would describe this specimen as *Shortia galacifolia*.)[137] As strange as it sounds, Gray's newfound familiarity with these European specimens made him the authority on North American flora.

In April 1840, Gray—tired of waiting for a paycheck from the University of Michigan—accepted Harvard's Fisher Professorship of Natural History, where he oversaw the botanical garden and taught botany. His specimens needed a home, and when Gray established the Harvard Herbarium in 1842, the second great American scientific institution was born. To stock his herbarium, Gray copied Torrey and either paid collectors or convinced organizations to obtain subscriptions to dispatch collectors.[138] These naturalists usually worked alone, sometimes under dangerous conditions. (David Douglas died suspiciously on the Island of Hawai'i in 1834;[139] Thomas Drummond died in Cuba in 1835; and Edward Frederick Leitner was killed and scalped by Native Americans in Florida in 1838.[140]) In addition, Gray and Torrey relied on freelance collectors (an odd occupation, but even in the old West, economics dictated livelihood). Following the drop in beaver prices in the early 1830s, the mountain men and voyageurs turned to other enterprises, and one way to use their hard-won knowledge of the West, retain their old lifestyle, and make money was to dig up plants. They would then send their specimens in bundles to Gray's and Torrey's colleague George Engelmann, in St. Louis.

Gray had met Engelmann in 1840. Engelmann was a German physician interested in botany who impressed Gray with his knowledge of the plants that Charles Geyer had collected for Joseph Nicollet along the Mississippi. In St. Louis, Engelmann was a point of contact for all scientists going west, and by 1841 Gray had a working agreement with Engelmann. When plants arrived in St. Louis "unassorted, unstudied, but full of new genera which had never before been described,"[141] Engelmann would organize and ship them to Harvard. Gray kept what he needed and sold duplicates—at a cost of about ten dollars each for Rocky Mountain plants, eight dollars for Texas plants—to defray the expenses of these expeditions.[142]

Torrey had firsthand experience with American plants, the best American herbarium, and a twenty-year record of working with the government dating back to Edwin James's collection from the 1819–20 Long Expedition. Torrey had also received plants from the 1820 Cass Expedition, led by Henry

Rowe Schoolcraft and D. B. Douglass to the Upper Mississippi River Basin, and from the U.S. Topographical Surveys, including specimens gathered by Nicollet and John Frémont, Nicollet's assistant and subsequently his successor. Frémont conducted four expeditions west, the first three while a member of the Topographical Survey, the fourth independently, on a trip to California that would eventually make him a wealthy U.S. senator and the first Republican presidential candidate. Frémont exhibited a characteristic of many public figures: the farther away from him you were, the better he appeared.[143]

In May 1828, Congress voted to send a naval expedition with civilian scientists to promote commerce and offer protection to the United States' hefty investments in the whaling and sealing industries.[144] After a number of unfortunate holdups, including the resignation of Gray as chief botanist, on August 18, 1838, the United States South Sea Exploring Expedition (U.S. Ex. Ex.) set sail under the command of Lieutenant Charles Wilkes. The nearly four-year journey took them south to Brazil, up the west coast of South America to Australia and New Zealand, to Antarctica, the Fiji Islands, Hawai'i, the Pacific Northwest, back across the Pacific to the Philippines and Borneo, then around the Cape of Good Hope, before arriving at New York on June 10, 1842.

The Wilkes Expedition played a critical role in development of nineteenth-century American science. Specimens collected during the expedition went on to form the foundation of the holdings of the Smithsonian Institution, established in 1846. The expedition's scientists, called "clam diggers" or "bug catchers" by Wilkes's sailors, explored about 280 islands, mostly in the Pacific, and mapped 800 miles of Oregon coast. They collected over sixty thousand plant and bird specimens. With no one else to turn to, and with hat in hand, in 1848 Wilkes appealed to Gray to study the expedition's tropical plants, which he did. These specimens were eventually sent to the National Herbarium at the Smithsonian, which Torrey helped establish.

In June 1846, the United States Senate approved the western boundary with Canada along the forty-ninth parallel. Two years later the Mexican boundary was finally settled when the Treaty of Guadalupe Hidalgo ceded to the United States what is now California and Utah, much of Arizona, and the remaining sections of New Mexico, Colorado, and Wyoming.[145] The American government sent expeditions to determine the exact positions of these international boundaries and supplemented their surveyors with geologists, botanists, zoologists, and ethnographers. As always, the motivation was commerce. After surveying the Grand Canyon in 1857, the Colorado Exploring (Ives) Expedition concluded, "Ours has been the first, and will doubtless be the last, party of whites to visit this profitless locality. It seems intended by nature that the

Colorado River, along the greater portion of its lonely and majestic way, shall be forever unvisited and undisturbed."[146]

In the early 1840s, Gray determined that "the interesting region (the most so in the world) is the high Rocky Mountains about the sources of the Platte, and thence south." This area was biologically less understood because it was off the main trade routes.[147] With this in mind, Torrey and Gray paid close attention to specimens from the Mexican Boundary Survey of 1849–51, as well as the Pacific Railroad Survey conducted along the thirty-fifth parallel (dubbed the Whipple Survey, after its commander, Lieutenant Amiel Weeks Whipple), through Albuquerque.[148] The botanists had advance notice of the railroad surveys. On March 19, 1853, Torrey wrote Gray: "Several expeditions across the country to the Pacific will probably be made this year. The routes for a rail road are to be surveyed.—& something will be done for Botany. We are keeping a good look out for our side."[149] By this time, the second volume of Torrey and Grays's *Flora* had been published, and as Torrey's biographer wrote, "Botany as a science, a logically organized science, one without amateurs, and with men of rare skill, philosophical judgment, and rare insight into the green life of the North American continent and its relations to other continents, was at last born."[150]

The Pacific Railroad Surveys, conducted from 1853 to 1855, consisted of four east–west surveys: the Northern Pacific Survey along the forty-seventh parallel from St. Paul, Minnesota, to Puget Sound; the Central Pacific Survey between the thirty-seventh and thirty-ninth parallel from St. Louis to San Francisco; and two Southern Pacific surveys, the Whipple Survey from Oklahoma to Los Angeles, and a second from Texas to San Diego. A fifth survey ran north–south along the Pacific Coast from San Diego to Seattle.

The purpose of these explorations (more like reconnaissances) was to find routes for transcontinental railroads. As with the border surveys, each team consisted of surveyors, scientists, and artists, and resulted in an immense body of data covering at least 400,000 square miles (1,000,000 km^2) of the United States west of the Mississippi River—essentially the Louisiana Purchase, the Spanish lands north of the Rio Grande, and the Pacific coast. These surveys were, ironically, authorized by then secretary of war Jefferson Davis (who would in 1860 become the president of the Confederate States of America) and conducted by Brevet Captain George B. McClellan (who would in 1860 organize the Union's Army of the Potomac). Findings were published in separate volumes, one per survey, by the United States War Department between 1855 and 1860, and contained extensive data on the natural history of the West as well as physical and cultural descriptions of the Native peoples encountered.

FIGURE 5. Joseph Hooker (sitting left, foreground) and Asa Gray (kneeling) at the camp at La Veta Pass (9,000 feet elevation) while in the Rocky Mountains during their 1877 expedition. John Muir would eventually join them in northern California. During this expedition, Gray and Hooker collected over 1,000 specimens (Turrill 1963, 166), and in 1880 Ferdinand Hayden published an essay they wrote on the geographical distribution of Rocky Mountain plants (Gray and Hooker 1880). Smithsonian Institution Archives, image #SIH MAH-19660. Both Harvard Herbarium Botany Library and Kew Gardens have copies of this photo. According to the Kew photo description, from left to right after Hooker and Gray are Mrs. Strachey, Mrs. Gray, Dr. Robert H. Lambourne, Major-General Richard Strachey, and Hayden. James Stenson, with hat, is standing between Dr. Lambourne and General Strachey.

It fell to Torrey to describe the plants from these surveys, and he enlisted Gray's assistance. Gray's work on these government surveys and his method of dispensing this information in limited-run, government-sponsored publications raised Joseph Hooker's derisive ire:

> What a pest, plague & nuisance are your official, semiofficial & unofficial Railway reports, surveys, &c. &c. &c. Your valuable researches are scattered beyond the power of anyone but yourself finding them. Who on earth is to keep in their heads or quote such a medley of books—double-paged, double titled & half finished as your Govt. vomits periodically into the great ocean of Scientific bibliography?[151]

Hooker's sarcasm aside, the regional flora of the West was generally known following the Pacific Railroad Survey reports.[152] Because biogeography never

much interested Torrey, Gray became "the first botanist ever to gain a simul-
taneous and accurate knowledge of the general outlines of the vegetation of
North America."[153] Nevertheless, there were shortcomings. Survey collectors
moved through areas quickly and overlooked species of small or otherwise
inconspicuous plants, species that bloomed in other seasons, or species located in inaccessible habitats. To correct these oversights, Torrey and Gray
sent their collectors to places such as remote mountain valleys, to stay many
years, collecting with a completeness the Pacific Railroad surveyors were un-
able to achieve.[154]

Gray also brought in Liberty Hyde Bailey. Bailey was born on March 15, 1858, in
South Haven, Michigan. In 1883 Gray sent a letter to young Bailey announcing
his new appointment at Harvard and his wish to find "for a year or two a man
who has the making of a botanist in him."[155] Six months later, Gray hired Bailey
for two years at a salary of $340 per year. At Harvard, Bailey worked as assistant
curator of the Herbarium and sorted Gray's western massive plant collection.
As plants were sorted, they were dispersed—one set went to the Missouri Botanical Garden, another to the Smithsonian.[156] In early 1885, Michigan Agricultural College needed a horticulturalist and offered the position to Bailey. Upon
hearing the news, Gray said: "But, Mr. Bailey, I thought you planned to be a
botanist." To which Bailey replied, "A horticulturalist needs to be a botanist."
And Gray retorted, "Yes, but he needs to be a horticulturalist, too." Another student, John Merle Coulter, thinking about Bailey's life away from East Coast intellectuals, continued the razzing: "You will never be heard from again."[157]
    Bailey was; his work in Michigan got him noticed. In early 1888, Cornell's
second president, Charles Kendall Adams, traveled to East Lansing to woo
Bailey. While they were meeting, Michigan State's president, Edwin Willits,
walked in. Recognizing Adams and quickly assessing his motives, Willits said,
"You're plowing with my heifer." The next day Bailey met with Willits and
informed him of Cornell's offer. Bailey emphasized that the issue was not
salary but instead centered on Cornell's wish to send him to Europe to learn
the latest European horticultural research. Willits responded, "In that event
you must go. You can do more for the advancement of Michigan Agricultural
College than if you stay here at what we can do for you."[158]
    Bailey thrived at Cornell. His first appointment—professor of practical
and experimental horticulture—led to promotions in 1903 to the director of
the Cornell Experiment Station and in 1904 to the dean of the New York State
College of Agriculture. Bailey edited the magazine *Country Life in America*,
and his collected publications while at Cornell are too numerous and varied
for even his most ardent admirers to track down. His many extramural activi-

ties included assuming the presidencies of the American Association for the Advancement of Science and the Botanical Society of America, and chairing Teddy Roosevelt's Country Life Commission. Bailey also co-wrote the Smith-Lever Act Extension of 1914, which formalized cooperation between land-grant universities and the U.S. Department of Agriculture to promote better agricultural extension.

Bailey was religious and, as with John Muir, saw God everywhere in nature. He was also an evolutionist who did not feel that evolution "destroyed faith except as it shattered doctrines and stripped religion of its 'non-essentials.'"[159] Bailey was a pro–Teddy Roosevelt Democrat, a "great, country-minded liberal."[160] His philosophical roots were aligned with the Transcendentalists, especially Thoreau, Emerson, and John Burroughs. Bailey felt rural life should be beautiful—bucolic and picturesque—and that nature's beauty should be preserved "along all steep and broken banks and sinuous water courses," as well as along roads and in rocky and untillable soils, where such preservation would not interfere with agricultural activities. He believed such "picturesqueness and beauty should portray color harmony and contrast form and perspective, and appeal to the mind."[161] Bailey sensed that nature could fill a spiritual void for urbanites[162] and believed: "Animals were in their places in the woods and they were all good. Whatever the human beings might have been in that community . . . the animals, at least, lived sensible lives."[163]

Bailey was also interested in forestry and in 1898 said:

> Forestry has not yet received adequate attention in the educational institutions of America. The forest is not only a stupendous crop which furnished fuel and lumber and material for a thousand trades, but it is a cover which conserves moisture, equalizes the distribution of water and protects the arable land. A large part of the country must always find its most profitable use in the growth of the forest cover, but the common intelligence upon the subject is so low that even wise legislation upon forestry matters is jeopardized.[164]

Bailey retired in early 1913. He simply locked his office door and left, following his philosophy of spending one-third of his life in training, one-third as a professional, and one-third in contemplation. This contemplation period lasted longer than he had planned; he died forty-one years later, in 1954. Bailey's intellectual descendants include Aldo Leopold,[165] Wendell Berry, and Wes Jackson.[166]

In 1846 Gray's and Torrey's good friend Joseph Henry became the first secretary of the Smithsonian. After the Philadelphia Academy of Natural Sciences and Gray's Herbarium at Harvard were established, the Smithsonian became the next great eastern scientific institution, and Henry's goal was to use it to acquire and disseminate new knowledge.[167] Under his guidance, this meant developing a system to receive, sort, distribute, and display the collections his institution was receiving, and Henry delegated this task to his assistant secretary, Spencer Fullerton Baird.[168]

Baird was born on February 3, 1823, in Reading, Pennsylvania. After a childhood immersed in natural history, Baird attended Dickinson College, where he was known as the "opossum hunter." Just before his graduation in 1840, Baird wrote a letter to John James Audubon describing two new species of flycatcher. When they finally met in February 1842, Audubon gave young Baird some tips on drawing birds. Audubon respected Baird so much, he named the last bird he discovered, in 1844, Baird's bunting (*Ammodramus bairdi*), known today as Baird's sparrow.

Baird's career was again boosted when, on a collecting trip in Vermont during the summer of 1847, the young professor of natural history, retained by his alma mater, encountered Congressman George Perkins Marsh, who would author the book *Man and Nature* and coin the term "conservationism in modern usage." Impressed with Baird's knowledge, Marsh later recommended that the new Smithsonian Institution hire him, and on July 5, 1850, the Board of Regents approved Henry's proposal to have Baird "take charge of the cabinet and to act as naturalist of the institution."[169]

The first collections that Baird incorporated into the Smithsonian were his own. At the time of his appointment, Baird's natural history collections filled two boxcars and included the following:

> About five hundred species of North American Birds, in skins, consisting of about twenty-five hundred specimens in various stages of age, sex, and season.
>
> A collection of the Reptiles and Fishes of the United States, at present contained in more than five hundred glass jars and numerous barrels, kegs, and tin vessels. Most of the species are represented by numerous specimens, amounting in certain cases to hundreds and even thousands of a single species.
>
> Skulls and Skeletons of many North American vertebrata, amounting to about six hundred specimens.
>
> A large collection of fossil bones from various caves in Pennsylvania and Virginia.[170]

Further, by hosting the National Museum of Natural History, the Smithsonian began receiving thousands of specimens from the federally sponsored surveys, including the border and railroad surveys in the 1850s and by the Geological Survey in the 1870s. Baird was meticulous and painstaking,[171] and

FIGURE 6. There was a time, in this case about 1890, when our nation's best biologists were popular enough to sell products, such as cigars. It was a man's world back then. Image is from the Smithsonian Institution Castle Collection, image #SI.1981.070, and used with permission.

his publications reveal the unprecedented depth of his knowledge of North American mammals, birds, reptiles, and amphibians.[172]

Similar to Torrey and Gray, Baird developed his own network of field-workers, providing them with collecting equipment and preservatives, including arsenic and cheap whiskey (having perhaps heard of Nuttall's problem, Baird treated his alcohol to make it unpalatable). He sent along items such as beads, knives, and mirrors for trading with Native Americans.[173] Collectors shipped their specimens directly to Baird, who would then sort, arrange, and classify them. Because Henry felt the Smithsonian should only retain type specimens, Baird distributed duplicates to other institutions.[174] Baird remained Henry's assistant for twenty-four years. The day after Henry's funeral, on May 17, 1878, the Board of Regents elected Baird the second secretary of the Smithsonian.

Baird's wife, Mary, was ill for much of her life. In the summer of 1863, seeking refuge from Washington's heat, the couple escaped to Woods Hole, Massachusetts. There, Baird first became aware of, and then concerned with, the decline of commercial fish stocks. In 1870 he sent a report to the chairman of the House Appropriations Committee requesting funding for "investigations into the subject of the food fishes of the Atlantic Coast, with a view of ascertaining what remedy can be applied toward securing the supply against its present rapid diminution." A month later he added, "I . . . suggest the appointment of a Fish Commissioner . . . whose duty it shall be to prosecute this investigation, and report upon these points to Congress." Baird helped draft a bill with these aims, which passed the House and Senate and was signed by President Grant in the spring of 1871. Grant then appointed Baird the United States Commissioner of Fish and Fisheries, without additional salary.[175]

After considering several locations for the Fish Commission's headquarters, in 1875 Baird settled on familiar Woods Hole. He justified his choice:

> The water is exceptionally pure and free from sediment, and . . . a strong tide, rushing through the Woods Hole passage, keeps the water in a state of healthy oxygenation especially favorable for biological research of every kind and description. The entire absence of sewage, owing to the remoteness of large towns, as well as the absence of large rivers tending to reduce the salinity of the water, constituted a strong argument in its favor.[176]

In 1883 Baird received a congressional allocation to construct a research vessel, the *Albatross*. In addition to his astounding breadth of knowledge, Baird had the capacity to attend to details, which he did in designing the *Albatross*. He ensured staterooms and laboratory space for the scientists and arranged for appropriate nets, traps, and dredges for oceanographic work. The *Albatross* remained in service for four decades, until she was retired in 1921.

Today, Spencer Baird is considered a transitional figure. His research was that of a nineteenth-century naturalist interested and expert in ornithology, mammalogy, herpetology, and ichthyology,[177] but at the same time he encouraged many of the best of the next generation's biologists, including Robert Kennicott, C. Hart Merriam, and Stephen A. Forbes. In fact, the Smithsonian attracted many young scientists to Washington, DC, and a group of them, led by the irreverent William Stimpson, founded a group called the Megatherium Club, named for a giant fossil sloth. This alliance, a proto–Cosmos Club, included the biologists Edward Drinker Cope, Ferdinand Hayden, Kennicott, Fielding Bradford Meek, Henry Ulke, and Henry Bryant. Stimpson lived in the Smithsonian Castle, which members dubbed the "Stimpsonian." They had a seminar program and invited biological dignitaries such as John Torrey and Louis Agassiz. Prototypical field biologists, members spent their weekdays hard at work, their weekends partying. It is said that after the young men serenaded Joseph Henry's daughters, he threw them out of the Castle.

Rob Kennicott.   Henry Ulke.
Wm Stimpson —   Henry Bryant.

FIGURE 7. Megatherium Club members William Stimpson, Robert Kennicott, Henry Ulke, and Henry Bryant hamming it up for the photographer. Smithsonian Institution Archives, image #SIA-SIA2008-0347. The Harvard Botany Libraries Photograph Collection also has a copy of this photograph.

Just prior to the Civil War, in 1859, two additional scientific institutions were founded. One, the Museum of Comparative Zoology at Harvard, where we have come to expect important institutions. The other, the Missouri Botanical Garden, at the gateway to the West.

Louis Agassiz came to the United States during the fall of 1846 from Switzerland. He had studied with Alexander von Humboldt, was considered the successor of the comparative anatomist Georges Cuvier, and dazzled as an original thinker on glaciation theory. Despite these accolades, Agassiz was cash-strapped. He had come to America to give a series of (paid) Lowell Lectures, planning to stay only a short time. However, the enthusiasm that Americans showed for both his lectures and his personality gradually convinced Agassiz he should stay. Further, he was appalled by the lack of quality scientific institutions in America and felt he could make a difference. Agassiz joined the faculty at Harvard and soon established, then dominated, the Lawrence

Scientific School.[178] In 1859 the Commonwealth of Massachusetts allocated $100,000 for Agassiz to build his Museum of Comparative Zoology (MCZ). Agassiz imparted the European-based master-and-apprentice model of education, unique in American higher education.[179]

In St. Louis, 1859 marked the opening of the famed Missouri Botanical Garden. British-born Henry Shaw had retired in 1839, after twenty years in the hardware business selling tools, cutlery, and other items to, among others, the mountain men. With $250,000 in his bank account, Shaw—aged forty, single, and living a relatively modest lifestyle—felt he had accumulated enough wealth and wished to enjoy it. In 1851 he visited the Duke of Devonshire's gardens at Chatsworth. Inspired, Shaw decided to construct a world-class botanical garden in St. Louis. Shaw sought advice from William Hooker, at Kew Gardens, who referred him to George Engelmann in St. Louis, who in turn referred him to Asa Gray. Hooker advised Shaw to build a library and a museum at the garden, to promote research. Shaw agreed, and when Engelmann was in Europe, he purchased books and secured the large herbarium of Johann Jakob Bernhardi.[180]

During the Civil War, in 1862, the northern states—comprising the United States—passed the Morrill Act. Shortly after, the Iowa legislature voted to accept its provisions, and Iowa State College became its land-grant institution, signaling a change in American education that would have deep ramifications for the fields of ecology and conservation biology. Iowa State would come to educate William Temple Hornaday and George Kruck Cherrie as undergraduates; count Charles Bessey among its faculty; train Arthur Carhart, the groundbreaking landscape architect for the Forest Service; and host, with Paul Errington as its head, the nation's first Cooperative Fish and Wildlife Research Unit. And if that was not enough, in 2000, two of its graduates, Mike Dombeck and Jim Furnish, set aside 58 million acres of road-less areas in America's national forests.

Following the Civil War, Cornell became New York's land-grant college. Located in Ithaca, the university founded in 1865 by Ezra Cornell and Andrew Dickson White opened in 1868; Cornell donated the land and the endowment, and White became its first president. White had attended Yale and found that the education there was "too much reciting by rote and too little real intercourse." He wanted a better educational offering. So did Ezra Cornell but differently. Cornell the man proclaimed at Cornell the university's inauguration, "I would found an institution where any person can find instruction in any study." Then, true to White's vision, Cornell University developed an elective system, where American students for the first time were allowed to choose their own curriculum. Cornell was also among the first universities to admit women, beginning in 1870.[181]

Yale's first museum specimens, collected in the eighteenth century, were gatherings of curiosities. Then, in 1802, Benjamin Silliman was appointed professor of chemistry and natural history and built a large and comprehensive mineral collection, which he used in his teaching. Silliman's collection became a public attraction, and by the first half of the nineteenth century, it established Yale as a major scientific center. Othniel Charles Marsh was attracted to Yale for its scientific reputation. When his uncle, the financier George Peabody, began to divest his fortune, Marsh persuaded him to donate. And with his gift of $150,000 in 1866, Yale's Peabody Museum of Natural History was born. Marsh was appointed both director of the museum and professor of paleontology, the first such academic appointment in America.[182] The first Peabody Museum building was opened to the public in 1876, and Marsh soon had it filled with dinosaur fossils. In 1924 the present-day Peabody Museum opened. Its Great Hall was designed to fit Marsh's dinosaurs, including the large *Apatosaurus*. In 1947 Rudolph F. Zallinger completed his 110-foot mural *The Age of Reptiles* on the Great Hall's south wall. Additions were built in 1959 and 1963, and today Yale's Peabody Museum holds approximately 13 million specimens.

At about the same time Cornell was established, New York's American Museum of Natural History was proposed. In 1869 one of Louis Agassiz's students, Albert Smith Bickmore, recommended to a group of men, including Theodore Roosevelt Sr., that a natural history museum be built in New York City. Three years later, they secured the property between West Seventy-Seventh and Eighty-First Streets, across from Central Park. Two U.S. presidents were involved in ceremonies attending to the museum's creation: in 1874 Ulysses S. Grant laid the cornerstone; three years later, Rutherford B. Hayes presided over its opening.

In the Midwest, Andrew Erkenbrecher, who had cofounded the Society for the Acclimatization of Birds, helped to create the Zoological Society of Cincinnati. On September 18, 1875, the Cincinnati Zoo and Botanical Garden opened its doors, becoming, after the Philadelphia Zoo, the nation's second-oldest zoo.[183] From its earliest days, passenger pigeons were exhibited at the Cincinnati Zoo.[184]

Just as Spencer Baird had impressed George Perkins Marsh in 1847, Clinton Hart Merriam impressed Baird in 1871. Young Merriam had always shown a great interest in animals. Around age five, he began hunting small birds and mammals with a bow and arrow; later he hunted with firearms, which his father replaced with newer and better models as he grew and became more proficient.[185] The mature Merriam was expert with both rifle and shotgun, an uncommon skill set. Once, in southern California, Merriam noticed two highwaymen pursuing him; he is said to have dismounted and shot them both.[186]

When Merriam was a teenager, his father, a U.S. congressman with wide interests (he had once corresponded with John Muir about Yosemite's glaciers[187]), allowed him to use a storeroom to dissect his bird and mammal specimens. His mother took exception to the stench, so Merriam was forced to apply basic tissue preservation skills (the use of corrosive sublimate and carbolic acid) to his taxidermy. He became proficient.[188] Not long after, his father took him to DC to meet Spencer Baird at the Smithsonian. Baird was stunned; most of Merriam's specimens were professional grade. Following Baird's recommendation, Merriam began advanced training with John Wallace, a New York City taxidermist whom Merriam later described as both "comical genius" and "reprobate." The following spring, Baird invited Merriam, then sixteen years old, to accompany Ferdinand V. Hayden's Geological and Geographical Expedition. In 1872 Hayden's survey traveled through Wyoming, Utah, Idaho, and Montana, with its primary focus on exploring for a second year[189] the newly established Yellowstone National Park. (The botanist John Merle Coulter was also on this expedition. He began as an assistant geologist and later became the expedition's botanist.[190] Moreover, another remarkable talent, Edward W. Nelson, was also on this expedition, as an assistant to the Megatherium Club member Edward Drinker Cope. Nelson would become chief of the Biological Survey, six years after Merriam retired.)

By 1872 the Army Corps of Topographical Engineers—begun in 1838 and led by unusually competent field men such as Stephen Long, Joseph Nicollet, and George Meade—had been disbanded (during the Civil War) as a unit and folded into the United States Army Corps of Engineers. Post–Civil War western surveys were conducted under the loosely organized United States Geological Survey. The Hayden Expedition—joined by Merriam, Coulter, and Nelson—was one of four explorations being conducted by the Geological Survey. As with the Pacific Railroad Surveys, the Geological Survey expeditions were directed by men with extensive field experience. Megatherium Club member Ferdinand Hayden led the Rocky Mountain Survey, begun in 1867. Clarence King led the survey of the fortieth parallel, also begun in 1867. (Should anyone think there were undo hardships on these expeditions, Wallace Stegner describes King's travel arrangements as follows: "At the Survey camp he was served by a black valet.[191] . . . [H]e possessed an apparently inexhaustible supply of fine wines, brandy, and cigars, and . . . his riding clothes . . . were made by London tailors out of snow-white deerskins dressed by Paiute squaws in the Carson Valley of Nevada."[192]) Lieutenant George M. Wheeler, a former Topographical engineer, led the surveys west of the 100th meridian. After King's survey was completed in 1872, another was added—in 1875 Major

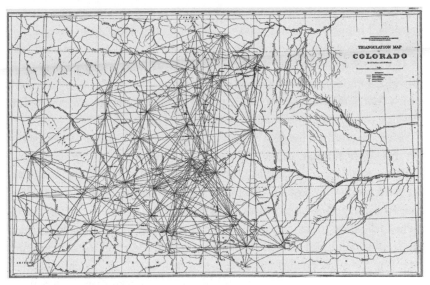

FIGURE 8. Ferdinand Hayden's triangulation map shows the amazing level of effort and detail these mid-nineteenth-century geographic surveys employed to learn the nation's topography. Reproduced with permission from the David Rumsey Map Collection (www.davidrumsey.com).

John Wesley Powell led the Second Division of the Rocky Mountain Survey to the Uinta Mountains. Powell, a one-armed veteran of the Civil War Battle of Shiloh, was by this time already famous for his earlier explorations of the Grand Canyon.

While the methods used by these four survey teams were similar, their administrative roots varied. King and Wheeler operated under the War Department, Hayden under the Department of the Interior. Through a congressional error, Powell was placed under Joseph Henry at the Smithsonian. When, in 1878, Powell's plan to consolidate these surveys was adopted and the U.S. Geological Survey was formalized, the army's role in western exploration came to an end.[193]

Young Merriam's "Report on the Mammals and Birds of the Expedition" appeared in 1873 as part of Hayden's Sixth Annual Report[194] and listed 33 species of mammals, 313 bird specimens, and 67 nests collected.[195] His job for Hayden completed, Merriam entered prep school at Williston Seminary in the fall of 1873, trained at Yale from 1874 to 1877, entered medical school at the College of Physicians and Surgeons at Columbia in 1877, received his MD degree in 1879, and became a practicing physician.

Despite pursuing this potentially lucrative career path, Merriam's avocation continued to be natural history. At Baird's invitation, Merriam spent the summer of 1875 working for the U.S. Fish Commission at Woods Hole.[196] At

Yale he worked in the new Peabody Museum and in 1877 produced his first important paper. His "Review of the Birds of Connecticut, with Remarks on Their Habits" was published in the *Transactions of the Connecticut Academy of Arts and Science*. Toward the end of his years at Yale, perhaps as a graduation gift, Merriam's father built a three-story museum on the family property, which Merriam used for specimen preparation and for storing his rapidly growing collection of birds, mammals, insects, and fossils.[197]

Merriam parlayed his dual interests in medicine and natural history by hiring on—again through Baird's influence—as a surgeon on the *Proteus*. During March and April 1883, this steamship with sails supported a seal hunting party along the coasts of Newfoundland and Labrador. While the sealers bagged a staggering 14,520 pelts, Merriam returned with about 120 specimens, delighting Baird—enough for him to request that Merriam overhaul the Smithsonian's seal collection. Baird also asked Merriam for sketches representing the seal's postures and behaviors in life, so that his chief taxidermist, William Temple Hornaday, and his assistants could produce realistic mounts.[198]

In early 1884, Merriam became the American Ornithologists' Union's chairman of the Committee on Migration and oversaw the first national bird count, which compiled data collected by volunteers across the country to inform species distributions and migration patterns.[199] Merriam wrote to his cousin, New York senator Warner Miller, about the possibility of receiving a federal appropriation for his committee's work, and in January 1885 the House passed a bill for the Bureau of Entomology, led by Charles Valentine Riley, to begin work on economic ornithology. The Bureau of Entomology had been formed within the Department of Agriculture as a federal response to an 1874 locust infestation that had begun in the West and spread to the South, destroying crops and forage. In 1877 farmers and ranchers persuaded the federal government to subsidize attempts to limit or eradicate these undesirable species, an ecological service recognized by the Entomological Commission Report of 1877, which promoted "the vertebrate enemies of insects . . . especially birds."[200]

In April 1885, Merriam was nominated to head the Division of Economic Ornithology. Baird seconded the motion, approved unanimously, and shortly after Merriam terminated his profitable but increasingly part-time medical practice to enter government service.[201] A year later, Merriam's political connections again served him well as Congress doubled his funding and expanded his responsibilities to include economic mammalogy. With this move, in spirit if not exactly in name, the Biological Survey was born (the Bureau of Biological Survey officially began twenty years later, on March 3, 1905).[202]

The legendary Merriam field method consisted of sending field parties out to collect mammals and other vertebrates in a region of interest; skins and skulls were taken from mammals, birds were stuffed, and reptiles and amphibians were fixed and stored in alcohol. In addition, biologists collected representative plants, took photographs of the habitats, and assembled a scientific report of the region. Most Survey biologists worked alone. Superiors such as Vernon Bailey and A. K. Fisher visited collectors at their field sites.[203] Bailey especially earned Merriam's respect. Born in Manchester, Michigan, in 1864, eighteen-year-old Bailey was living on a Minnesota farm when he sent Merriam 495 skins, 575 skulls, and 10 alcoholic specimens. A year later Merriam promoted Bailey to special field agent. In 1899 Bailey would marry Merriam's sister, Florence, herself an accomplished field biologist.[204]

At the time of Bailey's advancement, Merriam had thirty-four men and one business sending him specimens. In many instances, these collectors wrote asking for advice about, and criticism of, their preparation of specimens and were only incidentally concerned with remuneration.[205] Merriam's collectors included Ernest Thompson Seton, Nelson—later to be chief of the Biological Survey—and Frank M. Chapman, who would become curator of birds at the American Museum of Natural History and an outspoken critic of the plume trade. As with the Museum of Vertebrate Zoology's Annie Alexander, Merriam may have been his own best collector. His annual epic cross-country automobile trips over primitive roads to the West Coast in the spring, then back to the East Coast again in the fall, helped forge his legend.[206]

By the late nineteenth century, the Midwest had its land-grant institutions, its botanic garden and herbarium in St. Louis, and its zoo in Cincinnati. It would soon have its major research university in Chicago, and, following the 1893 World's Fair (named the Columbian Exposition, to celebrate the four hundredth anniversary of Columbus "discovering" America), it finally got its first great museum, the Field Museum of Natural History, in Chicago.[207] The first Field Museum, assembled for the World's Fair, was located in the Fine Arts Building in Jackson Park. The current Field Museum, made from marble and meant to last for eternity, was a part of the "Chicago Plan," whereby it, the Shedd Aquarium, the Adler Planetarium, and Soldier Field were built on Lake Shore Drive, rising from a "man-made peninsula" (i.e., landfill) claimed from Lake Michigan. Begun in 1913, pilings were driven into the lake bed and stone breakwaters were constructed. The area behind the breakwaters was filled with Chicago's garbage—ashes, tin cans, broken glass, and other waste—and covered with a thick layer of rich black topsoil gathered from an area west of the city that had not so long ago been tallgrass prairie. These magnificent midwestern public institutions of learning (and sport) were built on a garbage dump, which is as good an image as any for much of the world's intellectual progress. Herb Stoddard, with his Ringling Brothers experience, observed that when the Field Museum was moved, the "trainload of [taxidermied] large animals, including elephants, giraffes, and antelopes, riding in full view on flatcars up the Illinois Central tracks, looked like a giant circus moving up the lake shore."[208] (In 1899, in New York City, a similar train organized by Hornaday carried live animals to stock his Bronx Zoo.[209])

Back in Washington, DC, Henry Henshaw, Theodore Palmer, Waldo Lee McAtee, and Nelson formed Merriam's inner circle.[210] For a man who rode his bicycle to work, and what that implies today about a person's temperament, Merriam could be a difficult boss. When he became curator of zoology at Chicago's Field Museum, Wilfred Hudson Osgood, with the Survey from 1897 until 1909, complained that "the scientific men of the country look upon the Biological Survey staff as the best of its kind in the country. Yet [Merriam] cusses out everything done and must supervise every little detail and countenances no new conclusions whatever." Merriam had strong ideas about how the work of the Biological Survey should be conducted. He was not diplomatic nor did he mince words. To Merriam, an idiot was an idiot and that's what he called him. He was a storyteller and enjoyed dirty jokes. True to form, after once being painfully stung by a velvet ant, whenever Merriam saw one, he would stop his horse, dismount, and piss on it.[211]

But Osgood also saw Merriam "in toto":

> Few who knew him failed to realize that he was something beyond the ordinary; he swept people along with his own enthusiasms to such effect that only carping or jealous critics thought of him as egocentric. He was in fact very warm-hearted, very generous, and very sympathetic, but without his respect these qualities were not greatly exercised. He was not very tolerant of sloth, incompetence, or insubordination, but where these did not exist he was warmth itself. In the Biological Survey he occupied a pedestal, but he did not pose, for he detested insincerity. There was a certain indefinable magnetism about him which caused men of his own or even greater stature to be drawn to him quickly.[212]

Merriam dominated mammalogy in the late nineteenth century,[213] and when he retired, this last North American vertebrate class to be systematically investigated was broadly understood. What Gray and Torrey were to botany, and Baird was to ornithology, ichthyology, and herpetology, Merriam was to mammalogy. By 1886 Merriam had elevated mammalogy to a profession.[214] The challenge faced by mammalogists—why their species became the last to be understood—is that most mammals are, from a human perspective,

FIGURE 9. Biological Survey scientists at their U.S. National Museum office. From left to right: A. K. Fisher, Edward W. Nelson, Wilfred H. Osgood, and Vernon O. Bailey. Photo used courtesy of the U.S. Geological Survey–Patuxent Wildlife Research Center, Biological Survey files.

inconvenient. They tend to be nocturnal, small, secretive, or otherwise inaccessible. It was only after the invention of the cyclone small-mammal trap—the first in an ever-improving series of snap or deadfall traps—in the late 1880s that standardized collecting of cryptic species across broad geographic regions became possible.[215] With these new data rolling in, Merriam's group was soon publishing state-of-the-art taxonomic revisions based on a large series of specimens over a wide geographic range. Merriam used this information to construct primitive life-zone maps and to consider community-level interactions. For the first time, mammalian distributions and some form of ecosystem-level understanding of mammalian ranges became available.[216]

Unfortunately, though, for early mammalogists in general and Merriam in particular, mammals exhibit more variation in color, size, proportions, and skeletal characters than biologists had come to expect from either birds or fishes.[217] Therefore, when using these characters to define species, early mammalogists, especially Merriam, greatly overestimated their number. For

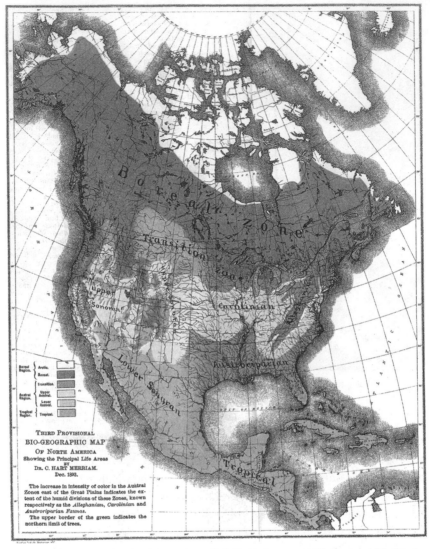

FIGURE 10. C. Hart Merriam's Life Zone Map, included in his 1893 report to Congress, published in 1894 (Merriam 1894).

example, in 1918, using skull characters, Merriam proposed eighty-six species and subspecies of grizzly and big brown bears in North America (today we recognize one species, brown bears [*Ursus arctos*], consisting of two subspecies, grizzly bears [*U. a. horribilis*] and Kodiak bears [*U. a. middendorffi*]).

Of all people, Teddy Roosevelt called out Merriam, and they debated this issue publicly beginning in 1897, both in the pages of the journal *Science* and on the floor of the Cosmos Club.[218] Roosevelt scored points when he noted,

"It seems to me the term species should express the fact of a wide and essential variation,"[219] and "if [Merriam] bases separate species upon characters no more important than those he employs, I firmly believe that he will find with each new locality which his collectors visit, he will get new 'species,' until he has a snarl of forty or fifty [wolves] for North America alone."[220] The problem of species identity plagued Merriam throughout his career. Much like John Torrey, Merriam gathered and organized facts, then published them without much considering their meaning.

In a letter to Osgood at the Field Museum, Roosevelt wrote:

> Merriam had it in him to be the greatest writer of faunal natural history in this country, or in the world. . . . He has [trundled] wheelbarrows with bricks instead. He has done capital work of [this] kind . . . but it would have been infinitely better if he had rigorously cut out a good portion of it, for instance, turning over the detail studies to young subordinates. . . . This would have been to work in the spirit of Cuvier, Humboldt, Darwin, Huxley. . . . [H]e ought to get out a book on the mammals of temperate America, which would last as long as our language lasts, which would be the best thing of the kind ever written about any continent.[221]

Despite these differences of opinion, and the gap between Merriam's potential and his reality, Roosevelt and Merriam were close friends. Roosevelt regularly sent Merriam notes of appreciation for his steady excellent work. There was an air of collaboration between the two, and Roosevelt enjoyed seeing Merriam's publications. Merriam's house was less than a mile north of the White House, so he would walk or bike down and visit Roosevelt during the latter's two terms in office.[222] When U.S. senator Henry Cabot Lodge wrote a letter to Roosevelt suggesting that Merriam, as a representative of the U.S. Biological Survey, was getting a little too enamored with Darwin's theory of evolution, Roosevelt wrote, "Now, I was a little disturbed at what you said to me about Hart Merriam. . . . On most matters I accept your judgment as much better than mine. On this you for the time being accept mine. The only two men in the country who rank with Merriam are [Alexander] Agassiz and [David Starr] Jordan."[223]

Merriam's conservation legacy is spotty by today's standards. During the 1880s, he severely criticized the indiscriminate killing of raptors, and the Survey's various bulletins on this issue gradually made an impression upon state conservation authorities, who were variously converting their old game departments into conservation commissions.[224] By the turn of the century, bird reservations had been established and the American Ornithologists' Union and the Audubon Society had been paying the salaries of the wardens. By

1909 the number of wardens had increased to fifty-one, and the private bird conservation organizations could no longer afford to pay them. To compensate these men and affirm their status as government employees, Merriam gave each a dollar a month. The situation embarrassed him, and before he retired, he arranged to have wardens paid twenty-five dollars a month.

Merriam did not extend his feelings about the conservation of birds to his chosen group, the mammals. For example, he felt the "great bulk of mammals are pests. Except the badgers, weasels, skunks, bats, moles, and shrews, very few of our mammals are of service to man. The bulk of the mammals, unlike the birds, are injurious to man, so that we have to fight them; and the first steps in fighting them is to find out what they are and what their life habits are."[225]

The Lacey Act of 1900 strained Merriam's resources. This law sought to limit animal imports; protect songbirds, game birds, and mammals; and enforce interstate trafficking in illegally killed game. These responsibilities fell to the Biological Survey, and they forced a shift in Survey priorities.[226] In 1905 Merriam lamented that Theodore Palmer, "the man who had been my best man, my right hand man for years, has for the last two years, given practically all his time to Lacey Act and related work, and that even with two other good assistants and a stenographer he still can not keep up with the work."[227]

Perhaps the most controversial practice of the Biological Survey was predator control. Farmers and ranchers increasingly lobbied for government intervention to deal with predators, rodents, and insects, and the Survey responded.[228] Merriam argued that the Survey's methodical extermination of wolves and coyotes through poisoning, trapping, and shooting was necessary because they were killing livestock. But it turned out that in addition to poisoning mammals, critics claimed people were eating the bait and also being killed.[229] The Survey's practice of setting poisoned bait traps would, half a century later, kindle the resignation of Olaus Murie, one of their superstars.[230]

During the first half of the first decade of the twentieth century, Merriam was at the apex of his career.[231] But by March 1909, Teddy Roosevelt was out of office, William Howard Taft was in, and there was a shift within the Biological Survey from basic science and biogeography to activities benefiting agriculture and business interests. Merriam had no stomach for this.[232] At the same time, Congress was threatening to shut down the Survey.[233] Merriam refused to give any ground by explaining or justifying the Bureau's activities to its congressional critics, who he felt had become an unnecessary hindrance to its work.[234] There was pushback, including a fake-news article that appeared in the *Washington Post* on January 24, 1907: "Frog science barred: No more experiments with the croakers." The article claimed that Merriam spent most of his congressional appropriation "in an effort to find a suitable home for frogs; that is, a place where they could propagate well, where the water was neither too cold nor too warm for them, and where the atmosphere was best for the improvement of their voices and incidentally where their legs would obtain the best development."[235] In 1910, fed up, Merriam resigned as chief of the Biological Survey.[236]

Merriam's transition into civilian life went uncommonly smoothly. Several of his friends, perhaps including Roosevelt, approached the recently widowed Mary W. Harriman about supporting Merriam. Merriam had organized the Harriman Expedition to Alaska in 1899 and was good friends with the family. In response to these overtures, the widow Harriman created a Harriman Fund to be administered by the Smithsonian Institution. The generous lifetime arrangement made Merriam a research associate at a salary of $5,000 per year, with an annual allocation of $7,500 to cover research expenses. In addition, Mrs. Harriman purchased his mammal collection, accumulated over four decades, for $10,000, which she then donated to the Smithsonian.

Merriam moved to California, where he pursued his ethnographic research interests. There, his good friends included John Burroughs and John Muir. According to his daughter, Merriam "was a great talker, enthusiastic in relating the many interesting and thrilling experiences he'd had. When he once started it was hard to get a word in edgewise. But once he met his match. John Muir and daughters were campt [sic] with us in Tuolumne Meadows. [Members of] the Sierra Club [were] nearby. None of these men were the silent type. It was fun listening to them—all such good friends—but not always agreeing!"[237]

Merriam's great friend Teddy Roosevelt needs no biographical sketch—there are dozens, with recent efforts typically focused on specific topics, issues, or relationships. As a child, Roosevelt was small, weak, and sickly; his mother dressed him like a girl.[238] As a young man, his detractors called him "Jane Dandy," "Li'l Pumpkin," or "our new Oscar Wilde." They made fun of his high voice and effeminate manner. In a homosexual reference, they said he was "given to sucking the knob of an ivory cane."[239] Roosevelt responded by lifting weights, boxing while he attended Harvard, and, for the rest of his life, identifying and meeting rigorous personal physical and intellectual challenges.

The bullying notwithstanding, Roosevelt grew up much like Baird and Merriam—shooting and mounting small birds and mammals to create a

museum-quality collection in anticipation of becoming a naturalist, specimen collector, or Survey biologist. Roosevelt recalled in his autobiography,
"When I entered college, I was devoted to out-of-doors natural history, and my
ambition was to be a scientific man of the Audubon, or Wilson, or Baird, or
[Elliott] Coues type, a man like Hart Merriam or Frank Chapman, or Hornaday, to-day."[240] But Harvard's late nineteenth-century curriculum, with its
emphasis on German-style benchtop biology, discouraged Roosevelt. He
wrote, ". . . at the time Harvard, and I suppose our other colleges, utterly
ignored the possibilities of the faunal naturalist, the outdoor naturalist and
observer of nature. They treated biology as purely a science of the laboratory
and the microscope, a science whose adherents were to spend their time in
the study of minute forms of marine life, or else in section-cutting and the
study of the tissues of the higher organisms under the microscope."[241] Following his father's death, Roosevelt gave up his dream of becoming a field
biologist, and in 1880—the fall after his graduation from Harvard—Roosevelt
and his brother Elliott went west, on a hunting trip they called their "Midwest
tramp."

During the first leg of their trip, Roosevelt found Illinois "shot out."[242] In
the town of Harvey, Roosevelt observed, "The farm people are pretty rough
but I like them very much. Like all rural Americans they are intensely independent; and indeed I don't wonder at their thinking us their equals, for
we are dressed about as badly as mortals could be, with our cropped heads,
unshaven faces, dirty gray shirts, still [dirtier] yellow trousers and cowhide
boots; moreover we can shoot as well as they can (or at least Elliott can) and
can stand as much fatigue."[243] During the second leg, near Carroll, Iowa, Roosevelt observed how "absolutely treeless" and "sparsely scattered over with
settlers" western Iowa was.[244] The third leg took them to one of the many western "sin cities"—Moorhead, Minnesota[245]—along the tracks of the Northern
Pacific Railroad. Although he had not seen bison, it was in Moorhead that
Roosevelt, for the first time, sensed the West he had only read about back
home, and he vowed to return.[246]

By the spring of 1882, young Roosevelt was back east, a New York State as-
semblyman leaning toward a political career. In April he wrote Elliott Coues
at the Smithsonian offering to donate the specimens he had secured as a boy,
a collection including both New World and Old Word species. Coues for-
warded Roosevelt's letter to Baird at the Smithsonian, and Baird and Roosevelt
began corresponding. Baird knew Roosevelt's father, who had helped establish the American Museum of Natural History. Roosevelt donated 622 bird
specimens to the Smithsonian, 46 from Europe and the Middle East, the rest

from his haunts in upstate New York. Baird valued Roosevelt's specimens, and they were incorporated into the Smithsonian's natural history collection.[247] Roosevelt similarly donated 125 specimens to the American Museum of Natural History,[248] including a snowy owl he had shot near Oyster Bay. This specimen became the pinnacle of Roosevelt's interest in natural history—a tourist attraction whose attractiveness grew with Roosevelt's legend.[249]

In 1883 the Northern Pacific Railroad was finally completed from Lake Superior to Puget Sound, and in August Roosevelt rode it on his second western trip. He was speechless at the beauty of the Badlands, but in this unfamiliar setting became unsure of himself.[250] "There were all kinds of things of which I was afraid at first, ranging from grizzly bears to 'mean' horses and gunfighters; but by acting as if I was not afraid I gradually ceased to be afraid."[251]

Roosevelt saw himself as a hunting cowboy. He invested in ranches near the Montana border, in Medora, North Dakota, and twenty-five miles to the north, a place he named Elkhorn Ranch.[252] Roosevelt bought cattle and used his ranches as base camps for hunting expeditions. As Stegner had described for Clarence King, Roosevelt had a buckskin suit made and shot a bison.[253] In the Bighorn Mountains, he shot a grizzly and a Bighorn sheep.[254] Besides his autobiography, he wrote three books—*Hunting Trips of a Ranchman, Ranch Life and the Hunting Trail*, and *The Wilderness Hunter*[255]—as well as many articles, either at his ranches or from his experiences at his ranches.

By 1886 the shine of being a rancher had worn off. Most of the bison herds were gone,[256] and their seasonal migrations were no longer a natural phenomenon. The winter of 1886–87 was the "Winter of the Blue Snow," when severe blizzards decimated cattle herds (the bison herds, had they been there, would have been fine).[257] By 1887 Roosevelt was done with ranching, although he continued to visit his Dakota properties until 1918, the year before he died.[258] And while he left the lifestyle, the Dakotas triggered his environmental awareness. It was one thing to have his father cofound the American Museum of Natural History; it was another to realize how the fragile resources of the arid West were being exploited. The response he envisioned had to be national to be effective.[259]

For the next fourteen years, as Roosevelt ascended the ranks of the Republican Party, he cultivated his network of natural historians. George Kruck Cherrie observed Roosevelt's nature and wrote, "The Colonel's friendly . . . and his almost boyish enthusiasm . . . won our confidence and loyalty at the outset. One of Roosevelt's most endearing qualities was his never-failing eagerness to hear of other men's aspirations and achievements, and to give full praise and credit wherever due."[260]

FIGURE 11. President Teddy Roosevelt and John Muir standing on rock at Glacier Point, Yosemite National Park, in May 1903; National Park Service photo RL012904.

Stirred as a child by the Transcendentalists Emerson and Thoreau, and by the adventure books of Thomas Mayne Reid,[261] Roosevelt was similarly moved by his contemporaries, John Burroughs and John Muir. But it was the scientists who inspired his awe—from his elder, Spencer Baird, to his colleagues C. Hart Merriam, William Temple Hornaday, and George Bird Grinnell. In turn, the people Roosevelt inspired included Gifford Pinchot, Roy Chapman Andrews, George Kruck Cherrie, and Liberty Hyde Bailey. And from among these people, Hornaday, Andrews, Cherrie, and Bailey were "pretty rough" midwestern men.

What It Meant

Guilds of like-minded people get lost in the face of disciplines fragmented into specialties. Gray, Baird, and Merriam all have biographies, but none of their biographers suggests that their subjects were components of an assemblage that described much of the vertebrate fauna and vascular plant flora of North America. As specimens were collected out west and sent back east, plants went to Gray at Harvard; fishes, amphibians, reptiles, birds, and some mammals went to Baird at the Smithsonian; and the rest of the mammals to Merriam at the Biological Survey. The various types of institutions (academic, museum, and government agency) supporting these men may have confounded pattern seekers. Nevertheless, much of everything in field biology that has happened since can be traced back to these men.

The Natural Historians

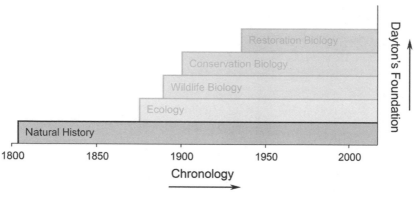

FIGURE A.

What fun it must have been to be an early naturalist discovering America's bio-diversity. There were hardships, of course, personal discomforts made worse in our imagination by living in today's climate-controlled environments. But the thrill of discovery must have surmounted everything. During our 1997–98 Ant-arctic trip, Joe Eastman and I discovered four new species of Southern Ocean fishes. I remember the weight—not excitement, at least not initially—of finding one in the pile of starfish, sponges, octopi, and familiar fishes phlumped from the trawl onto the deck of the icebreaker *Nathaniel B. Palmer*. I held the little black and orange snailfish in my hand and knew that in the entire half-billion-year history of life on Earth, I was the first human ever to have done so. The early naturalists must have had this experience on an almost daily basis. And, like me, many of these men were midwesterners.[1]

Bred in Their Bones

By the middle of the nineteenth century, the early primitive surveys of Lewis and Clark, Zebulon Pike, Stephen Long, and their contemporaries had given rise to the large institutional surveys of the Mexican and Northwestern borders, as well as the five Pacific Railroad Surveys and Hayden's state-based surveys. Each of these antebellum surveys was accompanied by the U.S. Cavalry and included geologists, botanists, zoologists, and ethnographers. As noted above, most of the plant specimens were sent to Asa Gray; the majority of the stuffed birds and mammals, as well as the pickled lizards, snakes, frogs, and fishes went to Spencer Baird.

Following the Civil War, the U.S. Geological Survey sponsored four western explorations. They continued Hayden's Survey, added King's and Wheeler's, then later Powell's. Baird was able to place promising young naturalists on these expeditions and cultivate their careers at the Smithsonian following their return. Hayden and Powell came from the Midwest, as did many of the men Baird favored.

Ferdinand V. Hayden was born on September 7, in either 1828 or 1829, in Westfield, Massachusetts. Hayden's parents had not been married. His mother was not interested in raising children, and his father was an alcoholic who had trouble holding a job. Hayden lived with relatives until 1841, when, at age thirteen, he was sent to live with his father's sister and her husband on a farm in Rochester, Ohio.[2]

In the fall of 1845, Hayden enrolled at Oberlin College, where he thrived in the classroom. However, fellow students noted his dominant personality trait of "crude ambition" and thought him uncultivated. They mocked him for falling in love with a new girl every week or two.[3] Hayden was not deterred and worked hard to improve himself. He read widely and joined one of Oberlin's literary groups. After graduation, Hayden began working with Jared Potter Kirtland at the Cleveland Medical School, where he distinguished himself in natural history.[4]

In 1853 Hayden became an assistant to the invertebrate paleontologist and Megatherium member Fielding Meek on an expedition to the Dakota Badlands. They traveled up the Missouri, where Hayden fell in love again, this time with the West. There, he made important professional contacts, especially among the men of the American Fur Company, including Gray's contact, George Engelmann, in St. Louis.[5] Meek's expedition was out only three weeks but collected hundreds of invertebrate and plant fossils, as well as scores of living plants, insects, amphibians, and reptiles.[6]

Hayden was an enthusiastic self-promoter, and in 1854 he convinced Baird to underwrite a two-year expedition to the Upper Missouri, including explorations along the Yellowstone River. He wrote, "It was wonderful country for a naturalist and geologist" and "with a Bag in one hand, Pick in the other, a Bottle of [collecting] Alcohol in my vest pocket, and with all [of] a gun to defend myself from Indians and Grizzly Bears . . . I could always return loaded down with rare and interesting things."[7] Hayden overwintered in Fort Pierre before continuing his explorations in 1855.

In 1856 and 1857, Hayden again went west, to Nebraska, as an adjunct to Lieutenant Gouverneur Kemble Warren's Corps of Topographical Engineers' survey. To assist his report writing and become better connected, Hayden, upon his return, moved to DC, where he joined Meek and the "Stimsonians."[8] In 1858 Hayden surveyed the Kansas Territory, and in 1859 he revisited the headwaters of the Yellowstone with a Corps survey team.[9]

In 1862 Hayden enlisted in the army as a surgeon, where he remained until the end of the Civil War. He then took a position as professor of geology and medicine at the University of Pennsylvania.[10] In the late summer of 1866, Hayden again explored Nebraska along the Niobrara and Little White Rivers, ending up at Fort Pierre, along the Missouri.[11]

Nebraska became a state on March 1, 1867, and its citizens were keen on a survey of its mineral and biological resources. On May 7, "U.S. Geologist" Hayden was awarded the $5,000 contract and conducted the survey. Among his conclusions were that Nebraskans should plant trees—that "most of the hardy northern trees may be cultivated on these western plains with entire success."[12] He wrote: "It is believed, also, that the planting of ten or fifteen acres of forest-trees on each quarter-section will have a most important effect on the climate, equalizing and increasing the moisture and adding greatly to the fertility of the soil."[13]

In 1869 Hayden shifted his interest to Colorado and continued his western explorations, funded through congressional allocations.[14] In 1871 Hayden got married and turned his attention to Yellowstone, where he would make his reputation. His 1871 survey (where, incongruously, Hayden's men played a baseball game against the troops at Fort Ellis[15]) is thought to have provided the evidence convincing Congress to designate Yellowstone as the nation's first national park. Hayden's Yellowstone surveys inadvertently created another success. The 1872 survey, through the appointments of several young assistants, deeply influenced the succeeding generation of American biologists.

Some quotes from the 1872 season offer insight into the thinking of that time. Hayden noted, first, that as one proceeds west, they "ascend barometrically, descend geologically."[16] Second, that on the Great Plains "it is not un-

FIGURE 12. The Hayden Survey team of 1872, which included C. Hart Merriam, John Coulter, and Edward W. Nelson, leaders of the next generation of American ecologists and wildlife biologists. Smithsonian Institution Archives, image #SIA-87-1882.

common for a river to be considerably larger toward its source than at its mouth."[17] Hayden continued this theme of aridity when he wrote, in a pattern of words as braided as the river: "Although the Platte River is never navigable at any season of the year, yet the area drained by it is immense, being nearly three hundred thousand square miles; and yet it is one of the minor branches of the Missouri River."[18]

Hayden's observations on native fish and game, as well as Native Americans, also mark his place and time. "Both Indian and buffalo have probably disappeared forever from these plains."[19] "Game, as antelope, elk, deer, bear, &c., was formerly abundant all over this region, and the experienced hunter might discover some even at this time; but all along the line of the railroad game of all kinds is fast disappearing."[20] And: "It is a curious fact that not a single trout has ever been taken in any of the branches of the North Platte, unless a few have been caught in the Sweetwater, while the branches of the South Platte are filled with them."[21]

About climate, Hayden wrote, "Facts . . . seem to sustain the popular persuasion in Kansas, that a climatic change is taking place, promoted by the spread of settlements westwardly, breaking up portions of the prairie soil, covering the earth with plants that shade the ground more than the short grasses; thus checking or modifying the reflection of heat from the earth's surface &c."[22] And "the general impression is, that all the lakes of the West are rising more or less."[23] Hayden also offered the contrary view to this prevailing idea that rain follows the plow. "A gentleman who has given much attention to meteorology writes me that he is not satisfied that settlements have sufficiently changed the surface of the country west of the Missouri to affect the climate. 'The increased rains,' he says, 'I apprehend are due to extra mundane or cosmic influences not yet understood.' "[24]

And finally, in considering the settlement of the Great Plains, Hayden wrote:

> At present one of the most important and interesting of the many questions relating to the great West is, How can the Plains be made more useful to man? And this, so far as it relates to agriculture, involves two other inquiries, as follows: How much of it can be irrigated to that extent required for the production of useful crops? And how much of the remainder can be profitably used as pastoral lands? The answers to these questions are of no small importance in the political economy of the nation, but, on the contrary, deserve and should receive the attention of our statesmen. That which adds to the material wealth and productive energies of a nation is of far more importance than that which simply represents value, although the latter often receives more attention than the former.[25]

Enter John Wesley Powell. Powell was born on March 24, 1834, in Mount Morris, New York. Powell's family moved west in typical settler pattern. He spent ages four to twelve in Jackson, Ohio, then up to age seventeen in South Grove, Wisconsin, before his family finally settled in Bonus Prairie, Illinois, just east of present-day Rockford.[26] Powell's college career was spotty; he spent time at Wheaton, Illinois College, and Oberlin.[27] During summers he took long collecting trips. In 1855, when Powell was twenty-one, he walked across Wisconsin. The following year, he floated the Mississippi from below St. Anthony Falls, Minnesota, to New Orleans. The year after, he floated the Ohio River from Pittsburgh to the Mississippi, ending up in St. Louis. In 1858 Powell first floated down the Illinois River, then rowed up the Mississippi River to Keokuk, where he continued his voyage up the Des Moines River into central Iowa.[28]

In 1858 Powell began teaching in Hennepin, Illinois, and was elected secretary of the Illinois Natural History Society, then located at Illinois State Normal University, near Bloomington.[29] In 1861 he enlisted as a private in the

Twentieth Illinois Volunteer Infantry, and a year later in the Battle of Shiloh he was wounded and lost his right arm.[30] Powell was discharged in 1865 with the rank of brevet lieutenant colonel and became a professor of geology at Illinois Wesleyan University. A year later, he assumed the same position back at Illinois State Normal.[31]

Over the next five years, Powell undertook four western surveys that would make him briefly famous and establish his career in government service. In 1867, with support from the Illinois Natural History Society, he explored the Rocky Mountains. Wallace Stegner asserted that at this time Powell "was not much of anything—not much of a scientist, not much of a teacher, not much of an explorer." But he brought to the problems of the West "science where [others] brought mythology, measurement where [others] brought rhetoric, and he brought an imaginative vigor . . . much better controlled and much closer to fact."[32] The next year, Powell again explored the Rockies, this time west of the continental divide, where he overwintered along the White River in Colorado. A year after, in 1869, Powell became famous for leading ten men in four boats down the Green River to the Colorado River, then down the Colorado through the Grand Canyon to the mouth of the Virgin River. In 1871 Powell led a second expedition down the Green and Colorado Rivers. Powell described his note taking: "My daily journal had been kept on long and narrow strips of brown paper, which were gathered into little volumes that were bound in sole leather in camp as they were completed."[33] By the end of these expeditions, which influenced his long bureaucratic career, Powell "would know the West as few men did, and understand its problems better than any."[34]

In 1870 Congress established the Geographical and Geological Survey of the Rocky Mountains and named Powell the lead of the Second Division of that unit, which explored west to the Uinta Mountains. Through a mistake by a congressional authorization committee, Powell was appointed through the Smithsonian Institution.[35] His Rocky Mountain survey ran for a decade.

In part because of Powell's insistence that the various individual geographical and geological surveys be centrally organized, in 1879 the United States Geological Survey was formed. After a bitter battle with Ferdinand Hayden, Clarence King became its first director, while Powell settled into the Smithsonian to work for Baird as the director of the Bureau of Ethnology.[36] In 1881 Powell left the Smithsonian to succeed King as director of the Geological Survey,[37] and in this capacity he organized the Irrigation Surveys, which later formed the basis for the Bureau of Reclamation. In 1888 Powell became the president of the American Association for the Advancement of Science.

Stegner described Powell:

F I G U R E 1 3 . The 1869 Powell Expedition down the Colorado, as commemorated in a United States postage stamp issued in 1969, marking the centennial of this voyage.

> He was one of the big men of the West, one of its foremost explorers, and in addition the father of half the scientific government bureaus. . . . Three [of these] government departments have had an incalculable influence in shaping the intermountain West, and through his sponsorship of them, Powell affected more people's lives than all but our greatest presidents have. . . . He was a bright and polished intelligence, a hotbed of ideas, an organizer, a martinet, a man of cool courage and the ability to plan.[38]

Just as importantly for Stegner, Powell was also one of those men of science with guts and integrity.[39]

Robert Kennicott was born on November 13, 1835, in New Orleans, where he spent his youth.[40] When he was an adolescent, Kennicott's family moved north of Chicago, to present-day Glenview, a hundred miles east of where John Wesley Powell was raised. When Kennicott was seventeen, he began collecting and cataloguing specimens for Baird at the Smithsonian. In 1855 he collected specimens for the Illinois Central Railroad Survey, including Kirtland's snake (*Clonophis kirtlandii*), a new species he discovered and described. During the Civil War, Kennicott worked at the Smithsonian. He became a member of the Megatherium Club and lived in the Castle. Kennicott wrote original descriptions of thirty-three new snake species and two genera, as well as five new mammal species.[41]

Kennicott did as much as anyone to demonstrate the benefits of Alaska to the United States. In 1859 he conducted a three-year exploration of the Yukon. Aided by the men and facilities of the Hudson's Bay Company, Kennicott collected plant and animal specimens and reported on important commercial factors such as arable lands, suitable harbors, and weather conditions. In 1864 Kennicott was at the Smithsonian when Western Union representatives approached Baird to request that he outfit an expedition to the Yukon to survey a route for a proposed telegraph wire to cross the Bering Sea from Alaska into Russia.[42] Baird was anxious to have collections from regions never explored by American biologists and quickly agreed. With his previous experience in the region, Kennicott was asked to lead the Smithsonian-based scientists, known as the Scientific Corps. This group included the artist George Maynard and the extraordinary naturalist William Healey Dall, who in 1884 had a subarctic species of sheep named for him by Edward W. Nelson.[43]

Any advantage Kennicott had in experience was offset by his unpredictable disposition. He often experienced periods of mania—staying awake and working for days at a time—punctuated by depression. His biographer suggested Kennicott took "Blue Mass" for his melancholy, the same medication Abraham Lincoln took for his. These pills were a concoction of elemental mercury, powdered licorice, rose leaves, honey, and chalk. As with much nineteenth-century medicine, the description of Blue Mass makes you wonder what was worse, the illness or the cure. Kennicott's bipolar tendencies made him emotionally unpredictable and prone to indecision, which did not endear him to either his Western Union superiors or his Smithsonian-based subordinates. Kennicott also suffered from a "weak heart" (bradycardia) and was taking the cardiac stimulant strychnine. He experienced fainting spells, including one in San Francisco prior to the Western Union survey, and one in Chicago four months before that.

Kennicott's group went to the Yukon during the late summer of 1865, over-wintered, and was still surveying for Western Union a year later when, perhaps having run out of strychnine, Kennicott walked away from his men, lay on a beach, and died; he was only thirty years old. An autopsy done in 2001 on his remarkably preserved body found no evidence of foul play, suicide, or poisoning, and concluded that, given his history of bradycardia, his heart simply stopped beating.[44] Today, Kennicott is remembered at his home in Glenwood, Illinois, called "The Grove," a National Historical Landmark. He is also memorialized at the institutions he furthered—the Smithsonian and the Chicago Academy of Science. But Kennicott is perhaps most remembered through the several species and Alaska landmarks named for him, including the now-abandoned Kennecott Copper Mines northeast of Valdez; the mines, although misspelled, are named for the adjacent Kennicott Glacier and, like The Grove, are a national historical landmark.

Robert Ridgway was born on July 2, 1850, in Mount Carmel, Illinois.[45] As a child he showed a keenness and aptitude for natural history, including collecting and drawing specimens—activities encouraged by his parents. In 1864, at the age of thirteen, Ridgway wrote to Washington, DC, seeking advice on the identification of a bird. He enclosed a full-sized color drawing, and Baird—reminiscent of his treatment of C. Hart Merriam—replied, identifying the bird and praising the boy's artistic abilities.[46]

In the spring of 1867, Baird arranged to have Ridgway appointed naturalist on Clarence King's Geological Survey of the fortieth parallel. Ridgway joined King's survey in May, after intensive training in specimen preparation at the Smithsonian. In an undertaking lasting nearly two years, Ridgway identified 262 species and collected 769 bird skins, most on the western slopes of the Sierra Nevada.[47]

Ridgway returned to DC and formally joined the Smithsonian in 1874, where he worked under another Baird protégé from the Midwest, curator George Brown Goode.[48] In 1880, at the age of thirty, Ridgway became the curator of ornithology ("Department of Birds"), and from 1866 until his death almost fifty years later, his title was "Curator of Birds."

Ridgway was an outstanding descriptive taxonomist. Working with the Smithsonian's collection of approximately fifty thousand bird skins, he unraveled the taxonomic relationships among North American species. As well, he continued collecting new specimens, making several trips around the United States and to Costa Rica.[49] During his lifetime, no ornithologist described more new taxa than Ridgway. He collaborated with Thomas M. Brewer and Baird on the five-volume *History of North American Birds* (three volumes on terrestrial birds published in 1874, which Ridgway illustrated, and two volumes on aquatic birds, published in 1884, which he helped write).[50] In its time, this work was the standard for North American ornithology.

Ridgway's crowning work was the monumental 6,000-page *The Birds of North and Middle America*, published by the Smithsonian in eleven volumes between 1901 and 1950. Begun in 1894 under the direction of Goode, it was technical, rigorous, and dry. Continuing the pattern of the *Manual* (and Baird's earlier *American Birds*), each volume featured an appendix of engraved outline drawings featuring generic characteristics. In 1919 Ridgway published the eighth installment of *Birds*, commonly known as *Bulletin 50*. Ridgway died in 1929, before he could finish this project. Following Ridgway's plan but doing

FIGURE 14. Photograph of Robert Ridgway in his office, August 1884. Ridgway's office was on the fifth floor of the south tower of the Smithsonian Building. In this room, Ridgway wrote over three hundred articles and drew the images for these publications. Smithsonian Institution Archives, image #91-2158.

his own writing, Herbert Friedmann of the Smithsonian completed the final volumes.[51]

Baird demanded from his biologists a precision of description, traceability through the literature, accumulation of empirical evidence (i.e., numerous specimens), and deductions drawn from facts. This was in opposition to the so-called European school (read "Louis Agassiz"), where credibility hinged on personal authority. Ridgway met Baird's expectations and, by following Baird's convention, became America's preeminent ornithologist.[52] He was a founding member of the American Ornithologists' Union (AOU) and was associate editor of the organization's journal, *The Auk*. Working within the AOU to reconcile the various systems of bird classification, Ridgway, Elliott Coues, William Brewster, Joel Asaph Allen, and Henry W. Henshaw formed a committee on nomenclature (taxonomy) and classification (relationships). In 1886 they released *The Code of Nomenclature and Check-List of North American Birds*—in essence a defined set of rules for the naming of birds to be described in the future and a consistent checklist.[53]

In 1899 Ridgway joined the famous Harriman Alaska Expedition, which Merriam organized. In addition to Merriam, John Muir, Louis Agassiz Fuertes, George Bird Grinnell, John Burroughs, William Dall, and a number of other scientists and artists supported by railroad magnate Edward Harriman joined in a two-month expedition to study the natural history of Alaska's coast.[54] Lewis and Clark's Voyage of Discovery opened America's nineteenth-century exploration; the Harriman Expedition closed it.

Seeking standards for accurate descriptions, Ridgway published two books that systematized color names for describing birds, *A Nomenclature of Colors for Naturalists* (1886)[55] and *Color Standards and Color Nomenclature* (1912).[56] Ornithologists continue to cite Ridgway's color studies and books. He was also an ambassador for the Smithsonian. Painfully shy, Ridgway was nevertheless an articulate public speaker and often served as the Smithsonian's spokesperson.[57]

Ridgway's boss George Brown Goode was born in New Albany, Indiana, on February 13, 1851.[58] His mother died a year and a half later, and by the time Goode was six, his father had remarried and moved the family to Amenia, New York. Goode was privately tutored until age fifteen, when he entered Wesleyan University in Middletown, Connecticut. He graduated in 1870 and attended Harvard, where he worked with Louis Agassiz. A year later, Goode returned to Wesleyan to direct its new natural history museum. The following summer, he volunteered for the U.S. Fish Commission in Eastport, Maine, where he met Spencer Baird. Goode spent the next five summers doing fieldwork for the Fish Commission and his winters either at Wesleyan or the Smithsonian. In 1877 he joined the Smithsonian full-time and became Baird's protégé. When the United States National Museum was established in 1879, Goode became its assistant director. In 1887 Goode became both the assistant secretary of the Smithsonian and the U.S. commissioner of Fish and Fisheries. He wrote over a hundred scientific papers in his short life; pneumonia killed him in the fall of 1896, at age forty-five. Before Goode died, he sent Teddy Roosevelt to meet William Temple Hornaday.

William Temple Hornaday was born outside Plainfield, Indiana, on December 1, 1854.[59] Three years later, Hornaday's family moved west to Iowa, where they bought a farm near Eddyville. In 1866 the farm failed and the family then moved to Knoxville, Iowa. Hornaday's mother died in January the following year and his father died three years later, making Hornaday an orphan at age fourteen. He rotated among family members in Iowa, Illinois, and Indiana, and at some point decided that under no circumstances would he become a farmer. Young Hornaday considered journalism and dentistry, then wisely decided to attend college before making his career choice.[60] At age fifteen, he entered Oskaloosa College, where he received his formal introduction to taxidermy.[61] During the spring of 1872, Hornaday left Oskaloosa and enrolled at Iowa State, where he fell under the influence of Charles Bessey, the twenty-seven-year-old chair of the Botany and Zoology Department. Bessey admired Hornaday's enthusiasm for natural history and wished to cultivate it. Bessey had a white pelican carcass. He removed volume 4 of Audubon's *Birds of America* from his shelf, turned to Plate 311, and instructed young Hornaday on how to bring the impression of life back to the dead bird.[62]

Hornaday described Bessey as "my guide, philosopher, and friend" and his tutelage under him as "a leading event of my life. . . . I have been very grateful for the fact that you started me exactly as you did. And did not discourage me by a long dose of anatomy, embryology, and gropings among the lower forms of life at a time when I wanted practical knowledge of beasts and birds and creeping things."[63] Just as importantly, as the pelican story suggests, Bessey strongly encouraged Hornaday's interests in taxidermy.[64]

In the 1870s, there was much to recommend the profession of taxidermy to a motivated teenager.[65] (Although had they known back then, there was much in terms of the effects of mercury and arsenic preservatives on an individual's life span to recommend avoiding taxidermy as a profession.) Across the United States, there was a large interest—part national curiosity/part civic pride—in building and stocking museums. At that time, museums usually provided the only opportunity for city dwellers to see wild animals. Further, with the land-grant colleges being established throughout the country, mounted specimens were critical for students learning zoology. Hornaday enjoyed taxidermy; collecting animals and preparing mounts fit his personality.

In 1872, as he walked the campus at Iowa State College, at a site just across Morrill Drive from the present University Library, Hornaday decided, "I will be a zoologist, I will be a museum-builder. I will fit myself to be a curator. I will learn taxidermy under the best living teachers—and I will become one of the best in that line. . . . This settles it! I will bring wild animals to the millions of people who cannot go to them!"[66] In 1916 Hornaday wrote a letter to Teddy Roosevelt describing the logic behind his transformational decision: "Shall I become a systematist and describer of species, and work for the scientific few? Or shall I devote my life to making animal and bird lore, and animal and bird specimens, available to the millions?"[67] Today, a brass plaque mounted on a small stone on the Iowa State University campus marks the spot of Hornaday's epiphany, placed there in 1926, on the occasion of Hornaday's retirement as director of the New York Zoological Park. It reads, in green patina, at a time when the evidence suggests ellipses were not understood: "This tablet commemorates the work of Dr. W. T. Hornaday for his contributions to zoology and conservation which have been of immeasurable benefit to America. It was on this campus as a student, June 1873, that. . . . 'I found myself . . . . .' "[68]

In early 1873, Bessey read to Hornaday an article about Ward's Natural Science Establishment, as the student worked in the university's museum (how times have changed). Ward's Establishment, located in Rochester, New York, was the leading company collecting preparing, and supplying specimens to the expanding museum and university markets. Impressed, Hornaday wrote to Henry Augustus Ward detailing his qualifications, with the qualification: "But my knowledge of the art is limited, and it is my wish and determination to make a first-class taxidermist." Ward, impressed nevertheless, replied with an offer, and when the term ended in November, Hornaday boarded a train for Rochester.[69]

Hornaday began at Ward's by cleaning up the taxidermists' messes. He was paid nine dollars per week; a third of that went to a senior taxidermist who was teaching him the trade. Hornaday soon mastered the basics and earned the favor of Ward. In early 1874, nineteen-year-old Hornaday convinced Ward to send him on a collecting trip, and after negotiations involving Hornaday's uncle, Hornaday left for Cuba and Key West. In Florida, Hornaday met Chet Jackson, who became his assistant. By the middle of March 1875, Hornaday and Jackson had collected twenty alligators, a large crocodile, a manatee skeleton, and several shark species.[70]

The large crocodile was dubbed "Ole Boss." Before shooting him, Hornaday observed and recorded the animal's postures and movements. Such natural details became characteristic of Hornaday's taxidermy. Ward sold Ole Boss to the Smithsonian for $250, which covered the costs of the entire expedition. In 1876 Ole Boss was prominently displayed at the Centennial Exposition in Philadelphia.[71]

During the fall of 1875, Hornaday and Jackson again traveled south. They spent eleven days in Barbados, two weeks in Trinidad, and nearly four months along the Orinoco River, where they collected piranhas, an electric eel, turtles, snakes (including a boa constrictor), alligators and caimans, birds (including macaws, ibises, toucans, and turkeys), anteaters, armadillos, sloths, ocelots, pumas, bats, and marmoset monkeys. They worked hard to get manatees; finally Jackson got a skeleton. Hornaday, his face sunburned and lips peeling, his mood cross and irritable, wrote: "I never worked as hard in my life for anything as for specimens while in the [Orinoco] Delta, and it makes me sick at heart to make me think of what we did not get."[72]

Late in 1876, Hornaday departed on a two-year, around-the-world trip. It was the beginning of the end for Hornaday's relationship with Ward. The two men traveled together for the first two months. They began their journey in England, traveled across the channel to the mainland and, from there, across North Africa. During their time together, Ward continuously niggled Hornaday, complained about his writing, and gave him tasks with few instructions and then criticized his performance. Over the course of their time together, other issues arose concerning salary, itinerary, expense funds, time frames, and demands for certain species.[73]

Hornaday arrived in Bombay on February 17, 1877. He found nothing in India to be as advertised, but he made friends quickly. Through these new friends, Hornaday arranged hunts in regions where game was plentiful. He was out two years and collected a diversity of specimens, including snakes, gavials, gazelles, buffalo, gaurs, axis deer, bears, langur monkeys, orangutans, two tigers, and three elephants. Broke, Hornaday returned to the United States via Australia, Singapore, Borneo, then across the Pacific. While in Singapore, Hornaday met Andrew Carnegie, who was traveling the world. They became friends and Carnegie adopted Hornaday's pet baby orangutan, "Old Man." Their friendship, and Carnegie's patronage, lasted until Carnegie's death in 1919.[74]

Ward was angry over the meager number of specimens Hornaday brought back, and in response Hornaday observed, "Ward never thanks anyone, or praises or compliments me in the least, and I am told others under him fare exactly the same."[75] Hornaday wrote a memoir of the expedition, which sold well and began to establish his reputation.[76] But what cemented Hornaday's status among professionals was the display he created with specimens from his expedition. Unimpressed by the shoddy workmanship and poor designs of the taxidermic displays in European museums, Hornaday conceived and then worked up "Fight in the Treetops," which illustrated a group of orangutans in a Borneo forest.[77] "Fight" debuted in August 1879 at the annual American

Association for the Advancement of Science meeting, where Hornaday complemented the exhibit by speaking on the natural history of orangutans. Impressed by Hornaday's talk and astounded by "Fight," George Brown Goode offered him a job. Hornaday deferred, honoring his contract with Ward's. Goode eventually got his man when, on March 16, 1882, Hornaday reported to the Smithsonian at a salary of $125 per month. As chief taxidermist of the National Museum, Hornaday became a clearinghouse for museums around the United States.[78]

Just as Hornaday was assuming his new position, the great bison herds were being extirpated from the Great Plains.[79] Like many, Hornaday realized the plight of bison, but his initial concern was not conservation but for the holdings of his museum—at that time, the Smithsonian did not have one respectable bison skin. Two months later, Goode and Baird sent Hornaday out to Miles City, Montana, to secure specimens. Baird requested that the commander of Fort Keogh provide a U.S. Army escort into what might be hostile territory. As well, Baird requested scouts, supplies, and camp equipment for the government-sponsored expedition.[80] When Hornaday arrived in Miles City, the smell of cattle manure permeated, a sign that the cattle industry had all but ended the roaming of the bison.[81] Hornaday's group managed to shoot two bison, which were shedding and therefore useless as skins. They also captured a calf, which they transported back to DC. The calf died and became a part of Hornaday's famous bison group at the National Museum.[82]

That same fall, Hornaday made a second bison expedition to Montana. By the third week in November, the group had killed twenty animals, and on December 12, just before they departed, Hornaday shot a large bull, which would also become a part of the National Museum's display. While preparing the hide, Hornaday found four older bullets embedded in its body.[83]

Never one to downplay his accomplishments, Hornaday announced that in his estimation, with the twenty-two skins, forty-four skulls, and eleven skeletons secured by his expedition,[84] the United States National Museum now had "the finest and most complete series of buffalo skins ever collected for a museum, and also the richest collection of skeletons and skulls."[85] Hornaday marked his hides with the letters SIBO (Smithsonian Institution Buffalo Outfit), mimicking the technique used by the old buffalo hunters.[86] On December 30, 1886, upon arriving in Minneapolis, Hornaday told reporters "that bison were indeed in a perilous position," and his expedition had been "made just in the nick of time."[87]

Throughout 1887, Hornaday worked on his six-animal bison exhibit.[88] One day the curtain surrounding the project parted, and a high, squeaky voice said, "Professor Goode said you wouldn't mind if I came down here and had a look, friend."[89] Teddy Roosevelt and Hornaday began talking about hunting, taxidermy, and the West, and they became fast friends.[90] In March the following year, Hornaday's bison exhibit opened at the National Museum, to wide acclaim.

Hornaday, as with many farm kids and artists, found that using his hands stimulated his brain.[91] While working on the bison exhibit, Hornaday conjured up the idea of a national zoo. He sent a proposal to Goode asking for $500 to kick-start the idea, which was put on hold because Baird was ill. Baird died two months later, and by December 31, 1887, with Goode's approval, a small wooden structure and primitive fence had been built to house Hornaday's few dozen animals, and his zoo was open to the public.[92] It was popular, and a month later Hornaday had fifty-eight animals. He then requested $15,000 to build a permanent learning-focused zoo.

On May 12, 1888, Goode appointed Hornaday curator of the Department of Living Animals of the United States National Museum. Hornaday's first funding request to Congress failed—he had attached the zoo appropriation as a rider to the District of Columbia Civil Sundry Appropriation bill. The second attempt was at first successful, and then it wasn't. The legislation to appropriate $200,000 to purchase land in Rock Creek Park for the National Zoological Park passed in March 1889, during the closing hours of the Fiftieth Congress, again as part of the Civil Sundry Appropriation bill. Grover Cleveland signed it during the last hours of his administration. In addition to allocating funding, the bill created a panel consisting of the secretary of the interior, the secretary of the Smithsonian Institution, and the president of the District of Columbia Board of Commissioners to oversee the construction of the zoo.[93]

While the National Zoo legislation was law, members of Congress successfully argued that a zoo fell outside the general bequest of James Smithson (the Brit whose funding had established the Smithsonian) and should instead be guided by the rules laid out by the Organic Act of 1878, which established a permanent government in the District of Columbia. A year later, Congress passed a bill that cut the zoo's operating budget in half and banned it from purchasing animals. Samuel Langley, who had succeeded Baird as director of the Smithsonian, would later write to his Board of Regents that this reduction in funding turned the zoo from "a scientific and national park" into "a local pleasure ground and menagerie."[94]

The 1890 bill also removed the zoo's governing panel and replaced it with the Smithsonian's Board of Regents, in essence putting Langley in charge. Langley, along with John Wesley Powell, decided that Hornaday did not have the executive experience and skills, nor did he have the academic credentials, to run a zoo.[95] After Langley further squeezed Hornaday, restricting his authority, Hornaday resigned.[96]

Hornaday then moved his family to Buffalo (which he attempted to get renamed to the more taxonomically correct "Bison"[97]) and began working as secretary of the Union Land Exchange.[98] While in Buffalo, Hornaday remained active in his former profession. He wrote a series of articles for *St. Nicholas* magazine, which provided the foundation for his 1904 book, *American Natural History*. In 1891 he published his popular how-to book, *Taxidermy and Zoölogical Collecting*. And in 1896, Hornaday published his one and only novel, *The Man Who Became a Savage.*[99]

On January 7, 1896, Hornaday received a letter from Dr. Henry Fairfield Osborn, the paleontologist jointly appointed by Columbia University and the American Museum of Natural History. (In 1905 Osborn would describe the most famous dinosaur species of all time, *T.* [*Tyrannosaurus*] *rex.*) The New York Zoological Society's Executive Committee felt that Hornaday was the only person in the country with the experience to build a large zoo in the country's largest city. Hornaday traveled to New York City, interviewed, and (Powell's assessments of his talents notwithstanding) was offered the job of director of the New York Zoological Park (the Bronx Zoo). Hornaday accepted and held this position until his retirement, thirty years later, in 1926.[100]

On November 8, 1899, the New York Zoological Park opened its doors to a large crowd. In the weeks before, railroad cattle cars filled with exotic animals funneled into the Bronx. This menagerie included pythons, an elephant tortoise, and birds such as ostriches, cassowaries, red-crested turacos, Impeyan pheasants, and harpy eagles. The assemblage also featured mammals such as kudus, waterbucks, Grévy's zebras, tapirs, capybaras, kangaroos, elephants, and orangutans. There was a colossal Alaskan brown bear and an African black rhino. Animals were sometimes acquired in unusual ways. As Hornaday recounted, a "beautiful great anteater [was] bought off a steamer from Venezuela for $40 and [will be] temporarily quartered in my office 'til next Monday, when the new Keeper's shed will be ready."[101]

Hornaday's Bronx Zoo offered 22 exhibits, featuring 843 animals representing 157 species.[102] About the amount and type of work it took to achieve his zoo, Hornaday exclaimed, "It does my heart good to wipe everything on my pants."[103]

It would have been enough had tiny Knoxville, Iowa, only generated William Temple Hornaday, but it also produced George Kruck Cherrie. As Cherrie wrote in the first sentence of his autobiography, "On the morning of August 22,

FIGURE 15. William Temple Hornaday, 1905, in his office at the Bronx Zoo. Photo used with permission of the Library of Congress, image LC-USZ62-102416.

1865, there was an increment of one in the population of the little village of Knoxville, Iowa. I was that increment."[104] Cherrie thus preceded ten-year-old Hornaday in Knoxville by a year. As Cherrie recalled, his first systematic work with birds was done in Knoxville "recording the dates of arrival and departure of the migrating birds" (the study of seasonality today known as phenology).[105] Cherrie followed Hornaday first to Iowa State then to Ward's. Again, from Cherrie's autobiography:

> At fifteen I entered the Iowa State College. . . . I took a course in mechanical engineering and specialized in mathematics, but spent much time in the little Natural History museum of the college. Leaving college I spent a year at Ward's Natural Science Establishment, Rochester, New York, where I met the late Carl Akeley and other men like W. T. Hornaday . . . leaders in museum development and Natural History work. . . .
>
> By 1888 I had begun to wander. I drifted down to Florida, thence to the West Indies. Since then I have gone on forty expeditions to the jungles of South and Central America and the wilds of little-known Turkestan, but have not always confined my activities to conventional Natural History explorations.[106]

Cherrie continued:

> After the World's Fair in 1892 [1893] I became Assistant Curator in charge of the Department of Birds at the Field Museum, Chicago. Wanderlust again seized me. From 1897 to 1899 I did natural history collecting for Lord Rothschild's Tring Museum and the British Museum. The Brooklyn Museum then put me on its staff. From there I joined the staff of the American Museum [of Natural History], and have recently been employed by the Field Museum.[107]

In the field, Cherrie carried a three-barreled gun, a sixteen-gauge double-barrel side-by-side on top, and a .30-30 rifle below. One sixteen-gauge barrel had a ".32-calibre shot cartridge" insert for small birds,[108] a common practice among museum collectors. Cherrie described the purpose of his profession: "The true naturalist is a hunter, not an assassin. He kills without ever being deliberately cruel. His aim is the advancement of science, and his success is achievement, not triumph. In my museum collecting I have secured well over one hundred thousand specimens of mammals, birds, reptiles and other animals."[109]

During the course of these expeditions—which Cherrie would describe as not particularly dangerous or arduous—he was attacked by pirañas,[110] narrowly avoided provoking a crocodile,[111] just missed being bitten by a fer-de-lance,[112] got attacked by a lesser anteater,[113] contracted a severe case of malaria,[114] and survived a flash flood.[115] What nature could not do, men almost did. Cherrie found himself in three separate gunfights, which in aggregate resulted in him killing two men and in turn being severely wounded. The bullet that hit him, shot by a man he killed by return fire, broke his arm and severed an artery. When Cherrie saw the severity of the wound, he thought to himself, using third-person, out-of-body thinking, "This ends Cherrie." Isolated in a remote jungle, he traveled overland several days then went by river to the hospital where his friend Dr. H. B. Parker was attending. Parker did not amputate because he considered the wound fatal. When Cherrie refused to die, Parker reduced the infection and removed the shattered bone fragments.

Cherrie spent five and a half months recovering, first at a hospital in South America, then in New York City.[116]

The memory of Cherrie would be faded today had he not been the chief naturalist on Teddy Roosevelt's River of Doubt Expedition in 1913–14.[117] Prior to the journey, Cherrie found his interview with Roosevelt much too brief for his satisfaction; he felt the ex-president should be made aware of his self-professed shortcomings. Cherrie recounted the conversation:

> "Colonel Roosevelt, I think that you should know a little bit about me before we start on this journey into the wilderness."
>
> "Well," said the Colonel without turning around, "What is it, Mr. Cherrie?"
>
> "I think," I said, "you should know that I occasionally drink."
>
> The Colonel went on writing. I waited—it seemed to me a very long time. Then he whirled around in his chair, looked me straight in the eye and said: "Cherrie, you say you drink?"
>
> "Yes," I replied, "I occasionally take a drink."
>
> "What do you drink?"
>
> And I replied, "Well, Colonel, that depends a good deal on what's available."
>
> "How much do you drink?"
>
> "All that I want."
>
> The Colonel shook a finger at me and said: "Cherrie, just keep right on drinking!"[118]

Teddy's son, Kermit Roosevelt, described Cherrie as a man

> of more than middle stature, sturdily but not heavily built, with both humor and grim purpose showing in his sunburnt features. "Many years of rainy seasons, and malaria's countless treasons"[119] had left their indelible marks upon his frame without affecting either his rugged strength or his powers of resistance. Father, as soon as we were alone, told me that he had at the outset reached the conclusion that no better companion could be found for an arduous voyage than Cherrie.[120]

Frank Alexander (Alex) Wetmore was born on June 18, 1886, in North Freedom, Wisconsin, outside Baraboo.[121] His father was a country doctor, his mother well-read, and consequently "his home [was] a place of books and ideas." Wetmore was always interested in birds, which he attributed to receiving a copy of Frank M. Chapman's *Handbook of Birds of Eastern North America*, when he was five years old. At thirteen, he published his first article in *Birdlore*: "My Experience with a Red-headed Woodpecker." As an adolescent, Wetmore collected and prepared bird skins, and by the time he was in high school, he was taking collecting trips. In 1908 Wetmore visited the American

FIGURE 16. River of Doubt Expedition members in Brazil. George Kruck Cherrie is bespectacled, third from the left, Kermit Roosevelt is seated on the ground in front of him, while Teddy Roosevelt is seated on the right. Photo used with permission of Houghton Library, Harvard University.

Museum of Natural History, where he met his hero, Frank Chapman, and saw his first European starlings across the street in Central Park.[122]

Wetmore attended the University of Kansas and worked at its Museum of Natural History. As an undergraduate, he was hired as a bird taxidermist at the Denver Museum of Natural History (now the Denver Museum of Nature and Science). In 1911 he left to serve as a field assistant on a U.S. Biological Survey expedition to the Aleutian Islands, led by Arthur Cleveland Bent. Wetmore earned his BA from the University of Kansas in 1912, then moved to George Washington University for his graduate work, where he received his MA in 1916, and his PhD in 1920. Now a DC insider, Wetmore held positions with the Biological Survey until 1924 and had a chance to get to know many prominent DC biologists, including Merriam, Ridgway,[123] and the young Ira Gabrielson.[124]

Wetmore banded ducks for studies on waterfowl management and investigated avian botulism in Utah's Bear River Marshes. In a separate study, he established the effects of ingesting lead shot on waterfowl.[125] In 1923 he led the Tanager Expedition, jointly sponsored by the Biological Survey and the Bernice Pauahi Bishop Museum, to the mid-Pacific.

Wetmore was tall, dignified, soft-spoken, and modest about his accomplishments. In 1925 he was appointed assistant secretary of the Smithsonian Institution, where he supervised the U.S. National Museum, the National Gallery of Art, and the National Zoo. As S. Dillon Ripley and James A. Steed pointed out, "Wetmore never pretended to enjoy administration. He admitted afterwards that he had always avoided administrative duty at the Biological Survey, either by leaving for the field or by sponsoring someone else for the post at issue. There are those who come to find administrative work interesting, an end in itself; to this group Wetmore clearly did not belong. . . . He believed the Smithsonian's way of doing business seemed least likely to hamper his research."[126]

While Wetmore researched fossil birds, he made his name studying living ones. With the Biological Survey, he did fieldwork throughout the United States and Canada, and in Puerto Rico, Mexico, South America, and the islands of the Pacific Ocean. Wetmore was an expert on the birds of Central and South America, and he authored several books describing them, including *A Check-List of the Fossil and Prehistoric Birds of North America and the West Indies* and *The Birds of the Republic of Panama*.[127] He focused on the Isthmus of Panama and established the Smithsonian's Canal Zone Biological Area (now the Smithsonian Tropical Research Institute). This region is a major flyway for birds migrating between North and South America. He would eventually describe 189 species or subspecies of birds, many from Central and South America,[128] and publish *The Migrations of Birds* in 1926.[129]

In the field, Wetmore always wore a khaki shirt and pants with a khaki tie. He expected camp to be laid out just so and therefore did it himself. As with Merriam and Cherrie, Wetmore was an expert shot. Once after he fired at a bird through dense vegetation, his field assistant Watson Perrygo asked if he had hit it. Wetmore replied, "I shot didn't I?"[130]

In 1945 Wetmore became the sixth secretary of the Smithsonian Institution. He followed the traditions of his predecessors by promoting laboratory and field research in natural history and anthropology. His challenges included managing the Smithsonian's collections, which were ever increasing and straining limited storage space; managing his staff, which was shorthanded, underpaid, and accommodated in substandard space; and properly funding his support structures.[131] Wetmore was able to modernize the administrative

FIGURE 17. Alexander Wetmore (*right*) and his field assistant, the taxidermist Watson Perrygo, collecting specimens at Stumpy Point, North Carolina. The two men were close friends. Among the Smithsonian staff, this image was known as the "Mutt and Jeff" photo. Smithsonian Institution Archives, image #SIA-81-13385.

organization at the Smithsonian, and through the Public Buildings Act of 1945, he created the conditions that twenty years later would result in the National Museum of American History, and a decade after that, the National Air and Space Museum.[132] Wetmore was secretary for seven years, stepping down in 1952 to create an opportunity for a younger person with a fresher perspective.

Wetmore was a member of National Geographic's board of trustees for almost forty years, and for thirty-five of those years, he was a part of the committee on research and exploration. The National Geographic Society published many of his articles and books, including *The Book of Birds* (1937, edited with Gilbert Grosvenor), *Song and Garden Birds of North America* (1964), and *Water, Prey, and Game Birds of North America* (1965). As Baird had before him, Wetmore held high standards. Wetmore was a member of the National Academy of Sciences and several other national and local organizations, including the Cosmos Club, of which he was president in 1938. Wetmore died on December 7, 1978, in Washington, DC. His eulogist concluded by saying what we all hope our eulogist will say about us: "He has gone from us in the accomplishment of grace."[133]

What It Meant

The archetypical legacy of field biology will come as a surprise to no one: field biologists stocked the great museums and herbariums of both the New and the Old Worlds. By the late 1870s, soon after the western surveys were consolidated and the U.S. Geological Survey was formed, most of the large organisms—anything that could be netted, trapped, shot, pulled, or dug up— had been collected and were known. Of course, new species continue to be described, but these tend to be found in remote places (including places such as redwood canopies) and tend to be small (think soil microbes and parasites) or visually or genetically cryptic.

Even with ongoing discoveries, it is fair to say that the age of organized scientific natural history in America began with Lewis and Clark's Voyage of Discovery and ended with the Harriman Expedition—explorations that neatly encompass the nineteenth century. By the end of this period, the natural history pieces—the species—were in place, and it was time to ask questions such as Why does this species occur there and not someplace else? How do these two species interact with each other? Does the hierarchy of biological organization (molecules to organelles to cells to tissues to organs to organisms) continue beyond individuals to include their groupings—populations and communities? In essence, we had the alphabet but didn't yet understand how the letters fit together to form words, phrases, poems, and novels. It was time for the ecologists.

The Ecologists

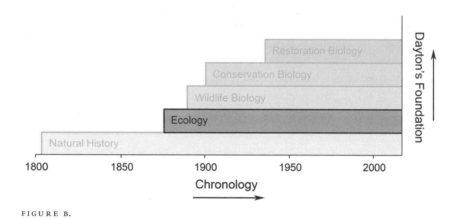

FIGURE B.

The earliest ecologists were Europeans, and it was a German, Ernst Heinrich Philipp August Haeckel[1] (the early champion of the "Out of Asia" theory of human origins), who, in 1866, coined the term "oecologie." Ecology's origins were, thus, unnatural in the sense that by the mid-nineteenth century, Europeans of one form or another had been doing pretty much whatever they wanted, not only to each other but also to the plants and animals they encountered, for a couple of millennia. At that time in the continent's history, there could be few ecological interactions that had not been affected by the hand of man.[2]

America's first ecologist was the Illinoisan Stephen A. Forbes. With his bird and fish food habits studies and his crop pest work, Forbes, more than anyone else at the time, examined interactions among species while pursuing questions in both basic and applied science. In the wake of Forbes, "ecology became self-conscious,"[3] and the four great early ecological schools defined by Frank Egerton—the Nebraska school of plant ecology, the Chicago school of plant ecology, the Chicago school of animal ecology, and the Wisconsin school of limnology—developed and grew.[4] It was no accident that they flourished in the Midwest. The science of ecology was developing just as land-grant universities were forming. Not long after, at Chicago, John Merle Coulter was determined to make ecology a central portion of the university's East Coast–provoking curriculum.

These big ecological schools are perhaps easier to recognize now than they were then. While there was antagonism—Chicago's Henry Chandler Cowles had no patience for Nebraska's Frederic Clements's hard-line views of succession—the schools as recognizable entities were secondary to the personalities of their protagonists. Representing Nebraska's plant ecology, Clements developed ecological techniques, studied prairies, took a hard-line approach to ecological succession, viewed ecological communities as superorganisms, and eventually collaborated with Victor Shelford, representing Chicago's animal ecology, to write the book *Bio-Ecology*. The Chicago plant ecologists taught us about the development and stabilization of sand dunes. Further, they showed that science is social, and therefore not always about Thomas Henry Huxley's nasty ugly little facts destroying beautiful hypotheses. Clements's hypothetical views of succession withstood valid fact-based attacks from the Chicago school for decades before finally succumbing.

The Chicago animal ecologists developed the technique of bringing complicated phenomena into the lab to isolate and study individual components. Shelford did it; so did Warder Clyde Allee. Ed Ricketts did it; so did Dick Bovbjerg. Shelford's collaboration with Clements on *Bio-Ecology* may have been torched by G. Evelyn Hutchinson's review in 1940, and as a result is almost forgotten today, but read *Bio-Ecology* and you will be stunned at the number and quality of ecological facts.

Edward Birge and Chancey Juday developed their limnological school in Wisconsin in parallel with the terrestrial plant and animal schools. For them, it was all about data and interpretation, over and over again, and they had no serious competition.

Bio-Ecology—the collaboration between Chicago's Shelford and Nebraska's Clements—signaled both the climax and the end of the run for these great ecological schools. While they emphasized ecology as organisms having interactions, with the natural history emphasis on organisms, Hutchinson's criticism of *Bio-Ecology* redefined ecology as interactions (such as energy flow and nutrient cycling) having organisms, with the heaviest emphasis on interactions (so called "modern ecology"). Both sides, of course, are correct, but it would take the environmental movement, with its focus on endangered and threatened species, to counter Hutchinson's perspective and bring respect back to organismal-based ecology.

An Illinois Original

Stephen A. Forbes was born on May 29, 1844, in Freeport, Illinois, between Rockford and the Mississippi River. Forbes spent his childhood learning the prairie plants and animals along the shores of the Pecatonica River.[5] When he was fourteen, Forbes attended a Lincoln-Douglas debate. While Lincoln was calm and measured, Stephen A. Douglas was hot and attempted to appeal to the crowd's emotion, calling opponents of slavery "black Republicans." The strategy backfired: Douglas was jeered, and when he complained of unfair treatment—since Lincoln had not been heckled—young Forbes, standing close to the stage, shouted out the reason, "Lincoln didn't use any such talk!"[6]

Forbes was quick-witted and well-read, so in the fall of 1860, at age sixteen, he attended Beloit College.[7] He did well, but his family could not afford for him to attend the spring semester, so Forbes worked until he and his older brother enlisted in the Union Army. The Forbes brothers were assigned to the Seventh Illinois Volunteer Cavalry Regiment. Except for the second half of 1862, when he was a Confederate prisoner, Forbes fought the entire Civil War and was discharged in November 1865.[8]

By early 1866, Forbes was in Chicago, at Rush Medical College, studying to be a physician. A year later, he completed his formal classroom requirements but had a change of heart, as he explained: "I have taken to botanizing lately and having learned to analyze, am making excursions after flowers almost every day. My little microscope helps and interests me very much."[9] Forbes then apprenticed with a former surgeon in his cavalry regiment, Dr. Thomas Rigg. Forbes spent time in South Pass, the Illinois portion of the Ozark Highlands. While collecting in the Cobden Hills, he came upon a fern he did not know (Goldie's shield fern, *Nephrodium goldieanum*), which he sent to George Vasey, the first president of the Illinois Natural History Survey. When Forbes's specimen arrived, Vasey was about to join John Wesley Powell's expedition to the Rocky Mountains.[10]

During this time, Forbes considered teaching: "In the company of my books and my own imaginations, I am thoroughly myself. . . . I find that I like teaching and am pretty well fitted for it."[11] During October 1869, Forbes became the principal of Union Public School in Benton, Illinois. He spent two years there and published his first scientific papers on the flora of southern

Illinois, including a description of the fern he had sent to Vasey. In 1871 he enrolled in Illinois State Normal University.[12]

At that time, Illinois State was among the top schools for teacher training. The faculty included Joseph Sewall, a Harvard alum who had studied with Asa Gray and Louis Agassiz. The school housed the museum of the Illinois State Natural History Society, consisting of sixty thousand specimens, including extensive botanical, zoological, and mineral collections—Robert Kennicott deposited specimens there.[13] As Forbes was completing his studies at Normal, the Natural History Society was in financial trouble and in the process of being rescued by a state appropriation. John Wesley Powell was the curator of the museum, but by then he was spending most of his time in DC finishing his survey of the Colorado River Valley. Powell finally resigned his curatorship and moved east in August 1872. His position was offered to Forbes, who, at age twenty-eight, accepted.[14]

Forbes was inspired by Agassiz's Anderson School of Natural History on Penikese Island and believed that "a general knowledge of the whole is of more value than a special knowledge of a very little."[15] Not long after assuming the curatorship at Normal, he began systematic field collections of freshwater invertebrates and fishes. During the spring and summer of 1875 and 1876, Forbes described a number of new taxa, especially crustaceans. He collected thoroughly across wet areas, including lakes, wetlands, rivers, streams, and springs, even sampling wells and drainpipes [16]

Forbes the generalist expanded his boyhood interest in birds to explore the controversy over whether birds assisted farmers, gardeners, and orchard owners by eating injurious insects or whether birds harmed these businesses by eating beneficial insects. Of course, the answer depended on many factors. Forbes believed that horticulturalists should have asked long ago, "How does Nature fight bugs?" and sought answers in that direction. To kick-start these studies, Forbes began shooting songbirds to examine stomach contents. He found variation in diet based on location, season, year, and age of the bird, as well as with competition across species.[17]

By the mid-1870s, the Normal Museum became so overloaded with specimens that Forbes suggested that it be converted into a laboratory for research and coursework in natural history and that the public natural history exhibits be transferred to a museum at the new statehouse in Springfield. The primary functions of the new laboratory would be to generate original research, disseminate scientific knowledge, and "translate the spirit and method of science with the work of the school."[18]

The result of this approach was a program whose progressive approach and accomplishments Forbes's biographer summarized as follows:

Forbes' and colleagues' work through the museum at Normal, the Summer School of Natural History, and the new state laboratory is a fine example of a uniquely enhanced natural history tradition. . . . Changes in the overall character of the old natural history included laboratory experience, critical examination of specimens by dissection and microscopy, active field excursions, study and course offerings for advanced and medical students and teachers, a state-supported biological laboratory with a prominent research function, and publishing of papers—all were evident in Illinois during the mid to late 1870s. This is the earliest occurrence of such a diverse set of activities for any state, and it illustrates Forbes' direct role and leadership during a major transition in the United States from natural history to a more professional ecologically based science.[19]

In August 1877 Forbes took an extended trip east to visit Boston (Harvard and its Museum of Comparative Zoology), Providence (Brown and its Jenks Museum), New Haven (Yale and its Peabody Museum), New York (the American Museum of Natural History), and Philadelphia (the Philadelphia Academy of Natural Science).[20] It was his first experience with big eastern universities and research museums. At this time, midwestern land-grant schools were just getting started, the University of Chicago was fifteen years away from educating its first students, and it would be a year after that before the Field Museum opened its doors.

Forbes's early fish and bird diet studies led to his interest in invertebrate populations, including economically destructive insect species. He found the most serious destruction to economically important crops came from insect species such as locusts and armyworms (actually the caterpillars of noctuid moths) that exhibited strongly oscillating numbers of individuals.[21]

More broadly, Forbes felt:

> Plants and animals are functionally connected and thus interdependent through their interactions. When the numbers, habits, and distribution of organisms are disturbed, these effects can spread through nature until readjustments occur naturally. Humans, by their presence, numbers, and actions, affect and disturb this "Natural system." . . . [I]f humans better understood how nature responded to disturbances, how the forces and processes by which such disturbances occur are reduced, aggravated, or even increased, [then] they might "by intelligent interference" avoid or reduce the negative effects of their actions. They might even enlist nature's help for the remedy and removal of these effects for society's benefit.[22]

By examining these food-based species interactions, and by using newly developed statistical techniques (for example, he based the number of samples he took on the variation he measured[23] and used a sophisticated coefficient of association[24]), Forbes became a leading, perhaps the leading, scientist during the rise of American ecology.[25] He also added to our vocabulary. In 1880 Forbes invented the term "microcosm," meaning a small world representing a much larger natural ecosystem.[26]

Late in 1881, faced with the task of sampling lakes in northeastern Illinois and southern Wisconsin and insecure about his skill set and equipment, Forbes wrote Spencer Baird asking if he could join the U.S. Commission of Fish and Fisheries' steamer *Fish Hawk* out of Woods Hole. Baird agreed, and Forbes accompanied a team of researchers from the U.S. National Museum, on a cruise designed to sample the Gulf Stream waters. Following the cruise, properly indoctrinated, Forbes stopped in Boston and loaded up with dredges, trawls, towing nets, and sounding lines for his midwestern freshwater surveys.[27]

In April 1882, state entomologist Cyrus Thomas—who had been the entomologist on Hayden's 1871 survey and who is credited with promoting the theory that the rain follows the plow—left Illinois to join Powell's group at the Smithsonian.[28] On July 3, Forbes was appointed to replace him. At about the same time, Forbes felt the State Laboratory of Natural History would be better located in Champaign, at the Illinois Industrial University, where the School of Natural History in the College of Natural Science was located. Others agreed. In 1884 Forbes was appointed to the School of Natural History at Illinois. Forbes transferred to Champaign: his collection consisting of thousands of specimens of fungi, fish, reptiles, insects, and other animals; his equipment and microscopes; as well as his library of 1,207 books and 3,856 pamphlets, papers, and periodicals. Things were happening fast for Forbes. In the middle of his move, he was recommended for the presidency of Iowa Agricultural College (Iowa State), and David Starr Jordan awarded him an honorary degree from Indiana University.[29]

The following year, the Illinois General Assembly formally approved the transfer of the state entomologist's office and the state laboratory to Champaign. It also approved a name change of the host university from Illinois Industrial to the University of Illinois. These moves did not, as detractors suggested, indicate a shift in the mission of the university to benefit "such branches of learning . . . related to agriculture and the medicine arts and military tactics" at the exclusion of "the scientific and classical studies."[30]

Forbes had strong opinions about the influence of native prairie borders on the productivity of agricultural fields. His data, collected at a time before the widespread use of industrial pesticides, suggested that removal of stalks, leaves, fallen and rotten fruit, decomposing wood, and native and invasive grasses and forbs from field edges and fence lines gave injurious insects fewer hiding places; therefore, Forbes sought clean fields. This approach is exactly the opposite of current thinking. The STRIPS (Science-based Trials of Row-crops Integrated with Prairie Strips) research team from Iowa State's Leopold Center for Sustainable Agriculture[31] is demonstrating that prairie strips increase diversity of native birds and insects, increase capacity for insect pollination, and decrease potential for erosion. Today, such prairie borders are encouraged.

The insecticides that Forbes used included arsenic-based compounds (Paris green and London purple) for "gnawing" species, and kerosene combined with soapy water for "sucking" species. The arsenicals had side effects on the plants being treated (and likely the humans applying them). These problems and many others were being studied at the newly formed agricultural experiment stations established at each land-grant institution under the Hatch Act, signed into law by President Grover Cleveland on March 2, 1887.[32] The University of Illinois Experiment Station opened in the spring of 1888,[33] the same year Forbes was elected dean of the College of Natural Sciences.

Forbes had an amazing and diverse energy. He slept little, had a big temper, and was agnostic, attending Unitarian services. He was a rapacious reader not only in the sciences but also in philosophy, history, politics, poetry, and French fiction. He enjoyed music. He smoked a cigar and pipe, and chewed tobacco, yet maintained his health through bicycling and jogging. In 1898 Forbes took a bicycle trip between Rockford and Freeport. He became the first president of the University of Illinois Golf Club.

Sports and fitness diversions aside, Forbes kept an eye on advancements in his field. In 1893–94, he wrote, "There is another division of biological science, little known to the general public by its name as yet, and but lately distinguished as a separate subject, which is now commonly called oecology. It is the science of the relations of living animals and plants to each other as living things and to their surroundings generally. . . ."[34] Forbes then went beyond this, into an area of debate that continues today. He believed an ecological view "must include the twentieth century man as its *dominant* species—dominant not . . . as simply the most abundant [predominant] . . . but dominant in the sense of dynamic ecology, as the most influential, the controlling or dominant member of his associate group."[35] If Forbes's judgment of talent was ever doubted, in 1914 he hired Victor E. Shelford as assistant professor of zoology and head of the new research facility of the state Laboratory of Natural History at Illinois.[36]

Forbes's aquatic studies continued mid-career. In August 1884, Baird convinced Forbes to investigate a massive fish kill in Lake Mendota, in Madison, Wisconsin, where Edward Birge taught zoology and was studying zooplankton

FIGURE 18. Stephen Forbes's glass-walled Entomological Laboratory at the University of Illinois, ca. 1900. Used with permission of the Illinois Natural History Survey, Prairie Research Institute, University of Illinois, image #1573.

(microscopic crustaceans).[37] Six years later, Forbes went west to Yellowstone to sample lakes and determine the nature and patterns of microcrustacean communities.[38] Forbes then stayed out west to work along the headwaters of the Missouri and Columbia Rivers, scouting locations for new trout hatcheries.[39]

In 1893 Forbes was appointed dean of the new College of Science at the University of Illinois. He also agreed to become director as well as assistant to the Board of Management of the U.S. Government Exhibits to the Columbian Exposition in Chicago.[40] While in Chicago, Forbes sent for his family, and during their visit his six-year-old son, Richard, developed diphtheria-like symptoms and died.[41] Forbes, perhaps blaming himself for Richard's death, never again mentioned his son's name in public.

Forbes viewed Illinois as a river state.[42] A year after the Columbian Exposition and now back in Champaign, Forbes requested $2,000 from the state board of trustees to establish a field station on the Illinois River to address the biology and ecology of a great river system "in interior North America." If funded, Forbes argued, the University of Illinois would be better equipped for freshwater research and study in ecology than any other university in the country. Forbes received his support and opened the station in Havana, an old riverboat and fishing town south of Peoria that supported the headquarters of the Illinois State Fish Commission. Optimistically, Forbes hoped the station would do for aquatic biology and fish culture what the agricultural experiment station was doing for agriculture and economic entomology.

The basic biological and chemical data collected at the Havana field station provided important baseline data for the changes that the Chicago Sanitary District and the Drainage Commission were about to impose on the Illinois River.[43] In 1900 Chicago reversed the flow of the Chicago River from its natural outlet in Lake Michigan to the LaSalle River and then into the Illinois River. The diverted channel comprised Lake Michigan water containing Chicago's municipal and stockyard sewage, as well as its industrial wastes.[44] To assess the effects of these pollutants, Forbes identified key indicator species of critical low oxygen conditions (hypoxia).[45]

In 1908 Forbes published *The Fishes of Illinois*,[46] and when the Illinois Natural History Survey was formed during the Illinois General Assembly of 1914–15, he was appointed its first chief.[47] Ten years later, he was elected to the National Academy of Sciences. Leland O. Howard nominated Forbes, pointing out "his great breadth of view as a zoologist and entomologist," and drove home his point by emphasizing, "It will be difficult if not impossible to point out a naturalist of his generation who was more original or broader or sounder." C. Hart Merriam may have voted on Forbes's nomination.[48]

The Nebraska School of Plant Ecology

William Temple Hornaday's Iowa State mentor, Charles Bessey, was born on May 21, 1845, on a farm outside Milton, Ohio.[49] Following the Civil War, he attended Michigan Agricultural College, graduating in 1869 with an interest in botany. He moved west and began teaching at Iowa State College of Agriculture in Ames. Bessey continued his studies while at Iowa State, and in 1872 received his master of science degree from the University of Michigan. He took time off to work with Asa Gray before receiving his doctorate at Iowa State in 1879. With Gray, Bessey worked on the taxonomy of vascular plants.

At Iowa State, Bessey established America's first microscope-based botanical laboratory. His dissertation resulted in the book *Botany for High Schools and Colleges* (1880), which made Bessey famous. He never studied in Europe or worked with Europeans, but his textbook represented the first American contribution to the latest advances in European experimental and microscopic botany, called "New Botany."[50] Agreeing with Teddy Roosevelt, Gray did not think too much of this newfangled microscopical and experimental approach, but praised Bessey's *Botany for High Schools and Colleges* as a "timely gift to American students of a good manual of vegetable anatomy and of the structure and classification of the lower cryptogamia, which was very much needed."[51]

Bessey was also interested in administration, and his time as acting president of Iowa State in 1882 and vice president from 1883 to 1884 led to the University of Nebraska hiring him with the triple duties of dean of the Industrial College, professor of botany and horticulture, and Nebraska state botanist. Shortly after, Bessey helped write the Hatch Act of 1887, establishing federally sponsored, university-based, agricultural experiment stations at land-grant colleges.[52]

Bessey flourished at Nebraska and from 1884 to 1907 built one of the nation's leading departments of botany, based on the foundations of new botany. He was a conscious innovator and continually warned of the "danger of falling into conservatism," of the "desirability of having a natural bias in favor of anything radical," and of the need "of having his brain and mind, theoretically and practically, all his life in a meristem state . . . guarding against the growth of much permanent tissue."[53] Bessey attributed his tendencies toward open-minded liberalism to a life in the Midwest and the freshness of its new land-grant colleges. He was young, full of potential, and at a new institution; he felt he could try just about anything.[54]

At Nebraska, Bessey's main interests were in new botany, applied botany, high school educational reform, and conservation, but his greatest accomplishment was the creation of a spirited, educationally progressive, professionally advanced, botanical "school," which flourished between 1886 and 1907, called "Sem. Bot." (Seminar in Botany).[55] As Ronald C. Tobey detailed, students in Sem. Bot. ranged from freshman to graduate students and included Frederic Clements and Roscoe Pound. Willa Cather was around, a friend of Pound's sisters.[56] Sem. Bot. functioned as an "invisible college," without official academic status but with initiation ceremonies and secretive rituals. Members met regularly to discuss the newest scientific discoveries, take field trips, and write research papers. They first established then conducted the Botanical Survey of Nebraska in the 1890s. Sem. Bot. originated the discipline of grassland ecology and developed methods of quantitative ecology.

When Congress established the Division of Agrostology (the study of grasses) in reaction to growing concern about the loss and degradation of grasslands due to poor farming practices, Bessey had the solution. Anticipating Aldo Leopold, he felt intelligent agricultural practices informed by science could rescue damaged land and get it on a path back to health, consistent with its natural direction. This process would, in time, come to be called succession.[57]

FIGURE 19. Charles Bessey working at his messy desk at the University of Nebraska. Photo used by permission of the University of Nebraska Library, Archives and Special Collections.

Bessey initially refused to accept the scientific credentials of ecology. He perceived it as dangerously close to "nature study" and considered it a fad—a waste of teaching and research time. His views did not, however, discourage his students' enthusiasm for ecology or their emphasis on field studies. Bessey's skepticism—sloppy science would not be tolerated—ensured his students, particularly Clements, fortified themselves by developing rigorous, quantified techniques.[58]

Frederic Edward Clements was born in Lincoln, Nebraska, on September 16, 1874. He attended the University of Nebraska, where he received his bachelor's degree in 1894 and his PhD four years later. In 1905 Clements was appointed full professor at Nebraska, but he left after two years to chair the University of Minnesota's Botany Department.[59] There, he met John Ernest Weaver. Weaver was born on May 5, 1884, in Villisca, Iowa. He earned his doctorate at the University of Minnesota in 1916. The year before, in 1915, he became assistant professor of botany at Nebraska, where he remained until his retirement in 1952, essentially extending the tradition of Bessey, and to a lesser extent Clements, into the second half of the twentieth century.

Weaver published over one hundred scientific papers and seventeen books, primarily on the vegetative features and ecological processes of prairies.[60] He was a slight, quiet, humble man, methodical in his preparation for classes and planning his research. Weaver created lists of things to do the next day, and he would check them off when completed. The result was solid progress and, as witnessed by his publication record, considerable productivity.[61]

Weaver was quirky. He never used a desk lamp; instead he wore a green visor to shield the glare from overhead lighting.[62] Weaver arranged to teach at 8:00 AM. He would arrive an hour early, teach for an hour, take a half-hour walk, then sit down to the rest of his day. Weaver wrote for two hours in the morning, ate lunch at a downtown diner at 11:30 to avoid the rush, then would come back to his office, take a nap, and work until 5:00 PM. He never left before and often took manuscripts home with him. Weaver's evenings were short. He went to bed early to think and kept a notebook on his nightstand.[63]

Weaver trained over one hundred MS and PhD students.[64] He paid for most of his lab's research using his own funds. He practiced what Dick Bovbjerg used to call "cigar box biology"—designing and building his own instruments and equipment out of spare parts, odd lengths of string, tape, tomato stakes, tin cans, rusty pipes, and so on. Weaver was demanding; some said too tough. He was distant from strangers and casual acquaintances, deeply warm with friends—"like a porcupine that liked to be cuddled."[65]

Weaver spent fifty years studying prairies. He examined just about every aspect—aboveground phenomena, belowground phenomena, single species investigations, soil effects, climate effects, disturbance, and succession. Weaver was similar to Clements in fussing over techniques, and their 1928 coauthored book, *Plant Ecology*,[66] is in many sections an updated version of Clements's 1905 book (below). Weaver differed from Clements, however, when it came to the relationship between scientific facts and theory. While Clements developed his theory of succession relatively early in his career and hung on to it in the face of fierce criticisms by Henry Cowles, Arthur Tansley, and Henry Gleason, Weaver was an inductive, facts-first thinker. For Weaver, theory derived from facts, and all theories were endangered by them,[67] which by contrast brings us back to Clements.

In 1917 Clements left the University of Minnesota to became an ecologist at the Carnegie Institution of Washington, in DC, where he was able to conduct his field and experimental ecological research year-round. He spent his summers at Carnegie's Alpine Laboratory, a research station on Pikes Peak, in Colorado, and his winters at field stations near Tucson, Arizona, or Santa Barbara, California.

FIGURE 20. John E. Weaver's studies of prairie plant root systems were unsurpassed. This image appears as fig. 52 on p. 147 of his 1954 book and demonstrates the development of the root system of little bluestem in excellent, good, and poor pastures.

Clements was quantitative—a stickler for techniques he felt would objectively assess important variables. In his 1905 classic, *Research Methods in Ecology*,[68] Clements pioneered the use of transects and quadrats for counting and measuring prairie plants.[69] This work was praised for its uncommon rigor, demanding attention to causality, and for its quantification—features characteristic of Bessey.

Clements viewed vegetation as dynamic. The vegetative unit, or organic entity, would change over time following disturbance, until it reached a "climax," a process termed succession. In *Research Methods*, Clements identi-

FIGURE 21. Frederic and Edith Clements examine their upturned car during a 1918 road trip. Metaphorically, this image depicts the fate of Clements's theory of vegetational succession at the hand of Henry Gleason. Reproduced with permission of the University of Wyoming's American Heritage Center. From the Edith S. and Frederic E. Clements Papers, image aho1678_000031.

fied both primary and secondary succession. Primary succession occurred through elevation, volcanic action, residuary soils, colluvial soils, alluvial soils, aeolian soils, and glacial soils. Secondary succession occurred through eroded soils, flooded soils, subsistence, land slips, drained or dried soils, animal agency, human agency, burned areas, lumbered areas, cultivation, drainage, and irrigation.

Clements equated succession to development and believed it was always progressive.[70] He envisioned successional processes driven by competition between plants with similar needs, and much of his experimental work was directed toward the analysis of different forms of plant competition. Clements, cleverly, used railroad right-of-ways as control sites, showing native, historical vegetational patterns. Not everyone agreed with Clements's views on succession, and his earliest, strongest opponents worked out of Chicago, on nearby Indiana sand dunes.

The Chicago School of Plant Ecology

The second great ecological school was also plant-based but had more distant origins. John Merle Coulter was born on November 20, 1851, in Ningbo, China, to missionary parents. Coulter's father died soon after, and his mother, a remarkable mathematician,[71] brought Coulter and his younger brother back to the United States, to Hanover, Indiana, where his mother's family lived. Coulter's maternal grandfather was on the faculty at Hanover College, a Presbyterian school founded in 1827—the oldest private college in the state.

Coulter led an unremarkable childhood until, as a teenager, he dove into a local swimming hole and severely cut his knee on the blade of a submerged ax. Despite first-rate medical attention, the wound became infected and Coulter spent a year in bed and another two on crutches. The injury affected him for the rest of his life. Coulter enrolled at Hanover in 1866, where he studied the classics. His junior year, Coulter came under the influence of Yale graduate Frank H. Bradley, a geologist. After Bradley mentioned to one of Hanover's trustees that geological evidence suggested an Earth substantially older than one based on a literal interpretation of the Bible, he resigned, likely encouraged by Hanover's administration.[72]

The following year Hanover hired another Yale-educated naturalist, Edward Thomson Nelson (no relation to Edward W. Nelson), to replace Bradley. Coulter and his younger brother Stanley were deeply influenced by Nelson's rigorous scientific mind. Coulter was curious about everything and under Nelson's influence became interested in botany. Together they catalogued much of the flora of Indiana and birthed the *Botanical Bulletin*, which a year later became the *Botanical Gazette* (the name change was to avoid confusion with the older *Bulletin of the Torrey Botanical Club*). Renamed the *International Journal of Plant Sciences* in 1991, Coulter's journal is still being published today.[73]

After leaving Hanover, Bradley landed on his feet. He was appointed chief geologist to Ferdinand Hayden's 1872 Geological and Geographical Survey of Yellowstone and brought Coulter along as his assistant.[74] During the trip, Coulter became the expedition botanist. C. Hart Merriam and Edward W. Nelson were also on this expedition.[75] In fact, at one point during the expedition, Bradley and Merriam were tasked with finding a route through the Tetons.[76]

Following this expedition, Coulter returned to Hanover with "1,200 or 1,500" mineral and fossil specimens—duplicates of material deposited in the Smithsonian—and the title of chair of Latin Language and Literature.[77] When Coulter went to Washington, DC, to work up his western specimens, he met Asa Gray. As the story goes, Coulter was working at the National Herbarium, his specimens spread across several tables, when an elderly man entered and started naming the plants. The young man was amazed and asked about him. "My name is Gray. I am interested in your work." Gray had read Coulter's report on the botany of the Hayden Expedition and became curious about the young botanist.[78] Their friendship would last until Gray's death in 1888. In 1874 Coulter published his early western findings as *Synopsis of the Flora of Colorado*, Miscellaneous Publication Number 4 of the Hayden Survey.

At Hanover, Coulter decided to make botany his profession, and in 1874, after the chair of the Natural History Department resigned, Coulter became the Ayers Professor of Natural History. A year later, Coulter published *Partial List of the Flora of Jefferson County, Indiana*—a compilation including 721 plant species, 367 genera, and 96 families.[79] Also in 1875 Coulter began the *Botanical Bulletin*, with his brother Stanley joining him as a coeditor a year later. In 1877 George Engelmann, Torrey's and Gray's longtime collaborator, contributed a paper, as did George Vasey, who had influenced Stephen Forbes. A year after that, Gray contributed an article, and soon Engelmann and Gray became frequent contributors, as did Charles Bessey.[80]

In 1879 Coulter accepted a chair at Wabash College, the second Presbyterian college (after Hanover) established in Indiana. During the summer of 1879, Coulter traveled to Harvard to take George Goodale's course "Experimental Vegetable Physiology."[81] Goodale was Gray's great friend and Coulter was impressed. Inspired by Goodale, Coulter returned to Wabash and taught "Analysis and Field Work" and "Lectures and Laboratory Work in Histology."[82] His Harvard connection established, Coulter bantered with Gray both in letters and publications, and Gray did not go lightly on Coulter, pushing him hard to realize his potential.[83]

Coulter became aware of Bessey sometime during the 1870s and was deeply impressed by his 1880 book, *Botany for High Schools and Colleges*.[84] Coulter allied himself with Bessey and new botany but did not abandon Gray's approach or perspective.[85] In 1885 Coulter went back to Harvard, in part to present Gray with a silver vase on the occasion of his seventy-fifth birthday. The horticulturist Liberty Hyde Bailey, fresh from a stint as Gray's assistant, was also there. Coulter stayed two months and wrote Bessey, "I have had a glorious time. [Mid-]Western botanists were as thick [at Harvard] as they have ever been."[86]

By the spring of 1891, the ichthyologist David Starr Jordan had left the presidency of Indiana University for the same position at Leland Stanford's new university in California, and the thirty-nine-year-old Coulter, Jordan's handpicked successor, took over.[87] It didn't work out. Either the size or the nature of the state job failed to appeal to Coulter. He enjoyed meeting with legislators but did not enjoy making endless requests for money. When he returned from these meetings, he said he always felt like washing his hands.[88]

By 1893 Coulter was the president of Lake Forest College in Chicago.[89] But the deep depression constrained Lake Forest's plans for expansion. A year later, when the newly established University of Chicago needed a botanist, Coulter began teaching there on Saturdays. Two and a half years later, in 1896, at age forty-four, Coulter left the presidency of Lake Forest College for a faculty position in botany at the University of Chicago.[90]

Until the Hull Biological Laboratories were built, botany at the University of Chicago was housed in two cramped rooms at the Walker Museum, so Coulter spent much of his time in the field.[91] Ecology was just developing as a science, and when Coulter brought in Henry Chandler Cowles from Oberlin, the two saw potential in studying plants in their natural habitats. Coulter had visited the Indiana Dunes as an instructor and guided Cowles's studies there. However, due to his bum knee, Coulter could not keep up with the rigors of sand dune fieldwork.[92]

Coulter did, however, keep up with botany by expanding the *Gazette* into an international journal. In 1886, along with Joseph Charles Arthur and Charles Reid Barnes, Coulter published the classic *Handbook of Plant Dissection* (otherwise known from the authors' initials as the "A.B.C. Book of Plant Dissection"). He had other responsibilities. In 1896 Coulter became the president of the Botanical Society of America. Within the next few years, Coulter published two successful elementary books, *Plant Relations: A First Book of Botany* (1899) and *Plant Structures: A Second Book of Botany* (1900).[93]

Coulter rescued young Henry Chandler Cowles as he was about to abandon Chicago.[94] Cowles was born on February 27, 1869, in Kensington, Connecticut.[95] He grew up cultivating plants and flowers[96] and loved to read. Cowles's favorite subjects were adventure, history, and natural history, and he subscribed to several magazines, including *The Young Oölogist* (the study of eggs). When Cowles was eleven, he began keeping a diary.[97] Reminiscent of Baird, Merriam, and Roosevelt, as a boy Cowles pretended to be a professional naturalist. He formed the "Kensington Weeding Society" and merged his "museum of curiosities" with his brother's to create "The Cowles National Museum."

Cowles's high school botany texts included Gray's *How Plants Grow* (1858);[98] later he purchased *Manual of Botany*. After high school, in 1889, Cowles attended Oberlin. Oberlin offered a true liberal education—it was the first to admit women and among the first to accept black students.[99] By the time Cowles reached Oberlin, he was already an "excellent field botanist . . . quick and sure and [with a] wide knowledge especially of vascular plants." At Oberlin, Albert Allen Wright honed Cowles's botanical and plant identification skills and taught him general and glacial geology. Wright appointed Cowles as his assistant in the botanical laboratory and paid him to mount plants in the Oberlin herbarium.[100] Cowles had an expansive mind and spent his senior

FIGURE 22. John Merle Coulter, seasoned and confident. From the Photographic Archive, Special Collections Research Center, University of Chicago Library. Image apf1-01953. http://photoarchive.lib.uchicago.edu/. Identifier: apf1-01953.

year focused on geology.[101] Wright suggested Cowles enter the newly formed University of Chicago, which he did in 1894; he attended for one semester before running out of money. Cowles then went to Gates College in northeastern Nebraska to teach the fall semester. Following a controversy over academic standards centered on plagiarism,[102] Cowles—who had tried to uphold Gates's integrity—left Nebraska for central Iowa. There, he combined his two interests of botany and geology to begin graduate work under Thomas Chamberlain. Cowles assumed a research assistantship as a special field assistant with the U.S. Geological Survey, where his job was to collect ancient plant material on the Southern Iowa Drift Plain.[103] It was during this time that Cowles failed Chamberlain's exam and Coulter intervened.

Cowles had been attending Coulter's Saturday botany lectures, and on January 4, 1896, the two men discussed Cowles's future. Coulter strongly encouraged Cowles to take up botany. After other talks and a field trip, Cowles joined the Botany Department,[104] where Coulter introduced him to the rudimentary field of ecology. Coulter saw ecology as the coming thing and wanted the "young, ambitious University of Chicago" to feature it. Doing so would distinguish Chicago from the big eastern colleges, which tended to view ecology through the lens of the back-to-nature movement and the "nature-fakers," and thus (showing the same skepticism as Bessey exhibited) pseudoscience.[105]

The Danish botanist Eugenius Warming deeply influenced Cowles. Warming felt that "natural forces were . . . at work to restore an equilibrium. Drifting dunes were being anchored by vegetation, lakes were being converted to swamps and bogs, while plant succession both on dunes and in bogs was moving on toward stabilized climax." Warming also felt that this succession "is in effect, part of a geologic process."[106]

The problem at Chicago—indeed, everywhere in the United States—was that the European ecologists published in their native languages. Charles J. Chamberlain recalled:

> None of us could read [Warming's] Danish except a Danish student, who would translate a couple of chapters, and the next day Coulter would give a wonderful lecture on Ecology. Cowles, with his superior knowledge of taxonomy and his geology, understood more than the rest of us, and became so interested that he studied Danish and, long before any translation appeared, could read the book in the original. . . . The treatment of such sand dunes as Warming knew, started [Cowles] on his study of the comparatively immense moving dunes south of the University.[107]

Ecology came hard on the heels of new botany, which Bessey had championed. Summarizing botany in the late nineteenth century, Coulter wrote:

> The geographical distribution of plants has received much attention for many years, but the earlier observers could do more than accumulate facts and outline general zones. With the development of plant physiology, it became possible to organize these facts upon a scientific basis, and this organization introduces us into the great modern field of ecology of which geographical distribution is a conspicuous part.[108]

Against this background, in late April 1896, Cowles went botanizing in Dune Park, Indiana. "This was my first experience in a sand dune country. We climbed up the wonderful piles of sand and saw acres and acres stretching up and down the lake, billowy like a prairie or vast drifts of snow. The sand dune flora is very characteristic and new to me."[109]

The Indiana Dunes stretch roughly twenty-five miles across the crescent-shaped southern shore of Lake Michigan. In the west, closest to Illinois and well back from the shore, are the 4,000-year-old Tolleston Dunes, completely stable and covered with vegetation. Eastward are younger dunes: Miller Dune, the easternmost, is still forming today. Shore dunes exist on the other Great Lakes, on some North American seacoasts, and in Europe, but none compare to those of Lake Michigan.[110]

Coulter suggested that Cowles investigate dune ecology, so he did. From 1896 to 1898, Cowles studied the dunes between Dune Park and Furnessville, Indiana, at all seasons of the year, and he visited dunes along the Michigan coastline in summer. Cowles took the train (the South Shore Railroad, nicknamed the Yellow Peril[111]) from his home to the dunes. Friendly engineers would stop between stations so he could unload his equipment near a study site. Cowles completed his doctoral dissertation, "An Ecological Study of the Sand Dune Flora of Northern Indiana," early in 1898, defended it in March, and got his degree in April.[112]

Cowles stayed at University of Chicago, where he advanced in rank from laboratory assistant in ecology (1897) to assistant (1898), associate (1901), instructor (1902), assistant professor (1907), associate professor (1911), and finally, professor of ecology (1915). In 1911 he published *Part Two: Ecology* of the three-volume *Textbook for Botany for Colleges and Universities*, nicknamed the "*Chicago Textbook.*"[113]

Cowles's professional life divided neatly into two parts: the years from 1896 to 1913, when research and publication took precedence, and the years 1914 to 1934, when his priorities became teaching, natural areas surveys, and conservation.[114] Cowles's wife shared that he did not "hunger and thirst" for recognition, and he preferred doing research and teaching to writing up results.[115]

Cowles studied sand dunes wherever he could, for example, in Holland and Belgium in 1905 and 1906 and in the United Kingdom during 1911. In 1916 he declared he had studied "nearly all the dunes of the world, having personally visited most of them and read about the others." Cowles's early emphases were professionalizing ecology and making sense of plant succession. When the American Association for the Advancement of Science organizers asked Cowles to summarize the state of ecological research of 1903, he called the field chaotic because ecologists could not agree on "fundamental principles or motives."[116] Cowles predicted that the great task ahead was to unravel "the mysteries of adaptation." He contrasted the ideas of Lamarck with Darwin and offered, "Many have taken for granted on one side or the other what ought to be a subject for profound investigation."[117]

In summarizing his own work on succession, Cowles concluded: "In the eastern United States the final formation is a mesophytic deciduous forest; farther to the north and in the Pacific states, it is coniferous forest; in the great belt from Texas to Saskatchewan the final formation is a prairie; and in the arid southwest it is a desert." In every case, the "ultimate plant formation is the most mesophytic which the climate is able to support in the region as a whole."[118]

Cowles observed that in the absence of fire or grazing, forests invade prairies, while in the presence of grazing and fire, trees retreat. He called this phenomenon the "prairie-forest conflict" and noted that prairie and forest have battled back and forth over many centuries as climate shifted. Cowles concluded that the "existence of a prairie in a given place in our middle-west may be due in part to factors operating at the present time and in part to the influence of cumulative soil factors operating through past centuries." He felt that the evolution of a prairie soil through the influence of prairie vegetation favors the persistence of the prairie, while the evolution of a forest soil through the influence of forest vegetation favors the persistence of the forest. Cowles reasoned that climate change "may disturb the prairie forest balance either way."[119]

The biggest challenge to Cowles's conclusions came from Frederic Clements. Five years younger than Cowles, Clements developed quantitative and statistical survey techniques, some of which, such as quadrat sampling, continue to be used today. Cowles taught these techniques but did not use them because he had doubts: "I do very little quadrat counting, even when I am working intensively on a small area. . . . I have more confidence in a subjective method than an arbitrary one like that quadrat. After all, one must select one's quadrats, and that brings in the subjective element."[120]

More importantly, Cowles's observations on vegetational succession contrasted with those of Clements. While Clements likened succession to the stages in the maturation of a plant and believed it was always progressive,[121] Cowles observed reversals—"retrogressive succession"—in the field. He also noted intermediary stages and many kinds of climax formations. In his review of Clements's *Plant Succession*, Cowles reiterated his differences with Clements. "The chapter in which the views of the author and the reviewer clash most sharply is the one on the direction of development." Clements "states positively that 'succession is inherently and inevitably progressive.' The reviewer is as positive in his opinion as ever that succession may be retrogressive as well as progressive, although of course progression is much more abundant and important." Cowles then introduced evidence from his own experience, as well as statements from the British ecologist Arthur Tansley, to demonstrate how Clements's theory "did not square with the facts,"[122] to little effect. So much for Huxley's notion of beautiful hypotheses destroyed by data.

In 1910 Tansley invited both Cowles and Clements to join the four-week International Phytogeographic Excursion (IPE) in the British Isles. Tansley felt such field meetings were necessary to understand an area's vegetation, and he organized the IPE with this in mind. Starting on August 1, 1911, the party visited sites in England, Scotland, and Ireland, ending at the British Association for the Advancement of Science meeting in Portsmouth. They ate and talked their way through the month. Cowles was "amazed," he wrote, at the "vast amount of wild country in densely-populated England—the extensive areas of the Broads, the sand dunes and salt marshes, the numerous heaths and moorlands."[123]

Two years later, the Americans reciprocated. The 1913 IPE began in New York City on Sunday, July 27, and concluded there in early October. Cowles and Clements co-organized, but Cowles was the principal. Often traveling in a private railway car, the scientists saw the most famous natural areas of the United States. After visiting New York City, the New York and Brooklyn Botanic Gardens, and Niagara Falls, they spent about a week in the Chicago area, including Indiana Dunes, journeyed to Lincoln, Nebraska, then cut across the Great Plains to Colorado Springs and Pikes Peak. Following a stop in Salt Lake City, they passed through Idaho on their way to Oregon and Washington, then traveled south to California, stopping at Yosemite National Park, California Redwoods Park, San Francisco, and the Salton Sea. In Arizona, they saw the Grand Canyon and stopped in Tucson, and then passed through Texas, New Orleans, and Washington, DC, on their way back to New York.

Hosting the IPE was the pinnacle of Cowles's academic career. Soon after, World War I broke out, making additional excursions impossible. After this time, Cowles produced little research. While he was always developing new ideas and insights, he shared them with classes and colleagues instead of publishing. He "devoted himself mainly to his students, rejoicing in their progress and in their subsequent accomplishments." He was "fair minded, modest, never at all aggressive, and never dogmatic, he preferred non-technical language, avoiding new terminology and the formation of any rigid system for his science." Above all, he "brought students to nature as well as to books." He believed thoroughly in out-of-doors teaching, making nearby vacant lands his ecological laboratory.[124]

During these field trips, when Cowles would stop to point out something, one student noted, "Observers were crowded, plants often small, and transitions

FIGURE 23. The International Phytogeographic Excursion at Sawyer, Michigan, on August 4, 1913. University of Chicago Library, Special Collections Research Center. Photo by George Elwood Nichols, image apf8-03365.

brisk. Few students had cameras. No one had a portable recorder, and accurate field notes were imperative. For survival we grouped spontaneously. In our trio, one got [a] view of the correct specimen, and even a scribble-sketch. Another got the Latin and common names. The third tried for specifics on soil, microclimate, and so forth."[125]

These were serious students and their approach extended to clothing. "We donned rather evident 'field' garb. Knickers, even jodhpurs. If you could afford them, knee-high laced leather boots—otherwise heavy socks to the knee." This footwear was "splendidly unsuited for scaling steep sandy blowouts or mucking around in swamps," while insect repellents were "limited and heavy on the citronella" and sunscreen "was nil." Students depended on "shade, hats, [and] long-sleeved shirts." Even with these discomforts, Cowles's field trips had an "indelible impact." "We were sensitized for life to our surrounding natural world. We became aware of its vulnerability and our responsibility for it."[126]

Botany 36 was a four-week field ecology course that took students into wilderness areas all over North America. Cowles originated Botany 36 in 1900–1901, and it became his signature course. For four weeks at a time, he took students to natural areas all over North America. While these were teaching expeditions, Cowles used them for research, and among his favorite destinations were the shores of Lakes Michigan, Superior, and Huron, which share a common geological history.[127]

Cowles planned his field trips carefully, checking unfamiliar territory in advance. In 1921, as he was preparing to visit California, he wrote to professor William F. Badè, an early member of the Sierra Club, who taught at the Pacific School of Religion in Berkeley, stating he wanted students to see every kind of vegetation, from lowland chaparral to alpine meadow. Cowles asked how to avoid the crowds at Yosemite National Park. As an alternative to Yosemite, he was considering Sequoia National Park, where the giant sequoias are "much more numerous and finer," and the Kern River Valley or Kings Canyon, which is "about as good as the Yosemite, and much wilder." Since he wanted to cover much territory in a short time, Cowles asked whether the group could "get about more speedily" in Yosemite or Sequoia. He wanted hotel recommendations, too, and closed with an apology for asking so much.[128]

When students signed up for Cowles's field ecology, they got an instruction sheet estimating their costs and indicating where to be and when. "It is imperative," these instructions stated, "each member of the party be a 'good sport,' putting up cheerfully with rain, hot or cold weather, mosquitoes, black flies, and with inadequate or unsatisfactory accommodations." Cowles wrote that he would not tolerate a "spirit of complaint" and reserved the right to "dismiss summarily from the class any who complain" or "fail to harmonize with the other members of the party." Students must have "stout tramping shoes, with leggings, if the shoes are low," he continued, and both sexes need trousers. The "most satisfactory field garb for women," he stated, "is a riding habit made of khaki or other suitable material." Students were expected to bring Gray's *Manual of Botany* with them and handbooks for the local area.[129]

According to Paul Sears, Cowles was a "superb field teacher," who used "his delicate gift of whimsy to make clear that field teaching is serious, responsible business." Too many field instructors point out anything that catches the eye and make impressively erudite comments, Sears continued, but they leave students "with no sense of the unity and interrelationship of what they have seen." Cowles always had clear notions of what students should see, learn, and take away from field trips.[130]

"As an interpreter and inspirer, [Cowles] excelled, writing and speaking with clarity and force," Sears wrote. "More than this, he had that priceless gift which is a sure sign of a feeling for proportion—a great sense of humor. This kept him from becoming doctrinaire, from ever relying on method as a substitute for judgment." Sears added that the University of Chicago's Botany

Department "brought in students from all over the world and subjected them, whatever their specialty, to [Cowles's] leavening influence." Cowles's daughter, Harriet, recalled, "My father was seldom happier than when he was on his way to visit a new place, accompanied, preferably, by a small group of interested Botany students." On some trips, the group traveled in large open seven-passenger Packard touring cars with cloth tops. "Often the roads were washed out, at which time we all got out and pushed. . . . Every so often the engines would overheat and I would be elected to run down the slope to get icy cold water from a mountain stream. I have since wondered why nothing cracked in those engines."[131]

By the 1920s, Cowles was a senior member of the Botany Department and increasingly preoccupied with departmental and service issues. In 1923 Coulter was seventy-two years old and Cowles began assuming his responsibilities, including serving as acting chair of the Department of Botany, where enrollments were decreasing.[132] After Coulter retired, Cowles became chairman.[133] In 1925 Cowles became coeditor of the *Botanical Gazette* and a year later became a trustee at Chicago's newly acquired Wychwood field station. Soon after, Cowles began to exhibit resting tremors, a symptom of Parkinson's disease. Cowles taught for as long as he could, but as the disease progressed, he increasingly mumbled his words and wrote his blackboard notes in smaller and smaller script. Students would stay after class to review his lantern slides without him.[134] In 1932 Cowles quit teaching. He died on September 12, 1939, at age seventy.[135] Afterward, his widow discarded or destroyed most of his papers, preserving only diaries, a few letters, and some family photos. Later she put most of the remainder of Cowles's papers in the trash, where they were hauled away.[136]

What Cowles could not do to Clements's successional ideas his Illini counterpart, Henry Allan Gleason, did, although it would take decades for Gleason's ideas to become mainstream. Gleason was born in Dalton City, Illinois, on January 2, 1882. He received his BS and MA from the University of Illinois in 1901 and 1904, respectively, and his PhD from Columbia in 1906. From 1906 to 1910, he returned to the University of Illinois, first as an instructor, then as an associate in botany. In 1910 he moved to the University of Michigan, first as an assistant, then as an associate professor of botany, where he stayed until 1919. That year he moved to the New York Botanical Garden, where he spent the next thirty years serving in various administrative roles and researching his interests in plant taxonomy.[137]

The New York Botanical Garden was established by the New York State legislature and signed into law on April 28, 1891.[138] Today, it comprises 250 acres of the Bronx, consisting of the gardens themselves, a conservancy, herbarium, and research library. During the late nineteenth century, the idea for a botanical garden in New York was in the air, and the New York Botanical Garden resulted from a combination of private and public initiatives begun by the Torrey Botanical Club facilitated by the New York City Parks Commission. To stock the garden, trees and shrubs were bought or donated. To support its research mission, during the spring of 1896, Columbia University agreed to lend the garden its exceptional herbarium. The garden also purchased the mycological collection of J. B. Ellis, from New Jersey.[139] Nathaniel Lord Britton was a founder and its first director. While stumping for the Botanical Garden's establishment, he offered: "No city in the world can boast of such a beautiful and desirable site for a botanic garden as New York. . . . The Bronx Park, although now so little known to New Yorkers, is destined to be one of the most beautiful and popular of the public parks in the world."[140] According to Sharon Kingsland, the establishment of the New York Botanical Garden in 1891 signaled the beginning of American ecology,[141] a remarkable assertion given all that we have been considering here.

While young and in the Midwest, Gleason pursued ecological questions. He was deeply influenced by the theoretical work of Cowles. Unlike Cowles, however, he held the quantitative techniques of Forbes and Clements in high regard.[142] Gleason published four papers on quantitative ecology during the 1920s.[143] In them, he addressed problems associated with statistical methods, such as the necessity of locating quadrats randomly (Cowles's objection to their use), the effects of quadrat size and number on results, the tendency of older plant assemblages toward randomness, frequency distributions, and species-area curves.[144] Gleason was "the first . . . to study quantitatively the distribution of individual plants."[145] Gleason's results, especially in the area of interspecific associations, led him to challenge Clements's theories of plant assemblages and the natural process of attaining these assemblages, through succession.

Clements viewed plant communities as tightly interacting, interwoven assemblages—as superorganisms. It was as if individuals did not exist as independent units but were instead marching in lockstep with conspecifics and other species—as if the structure of life ran from cell to tissue to organ to organism to community, and that an organism could no more exist outside of its community than a brain could outside of its skull. Gleason, in contrast, felt that species distributed themselves independently, according to their own physiological, biochemical, and biological tolerances—the "individualistic hypothesis." Further, while Clements viewed vegetational succession as a closed system, proceeding through a series of ordered stages toward a fixed climax, Gleason viewed succession as an open process, the by-product of the responses of individual plants to their immediate surroundings, both abiotic and biotic.[146]

Gleason first challenged Clements in 1917, then drove home the attack in 1926.[147] To no avail. George Nichols considered Gleason's individualistic hypothesis in August of that year at the International Congress of Plant Sciences, and while he agreed with Gleason's facts, he disagreed with his conclusions. Nichols felt that plant associations were definable entities, similar to species, and that intermediates were similar to hybrid or incipient species.[148] Throughout the 1930s and most of the 1940s, Gleason's ideas continued to be ignored, despite the fact Gleason published a follow-up paper in 1939,[149] with the same title as his 1926 paper.

Gleason's individualistic hypothesis finally began to gather traction in the late 1940s and early 1950s, and was fully embraced by the late 1960s. John Curtis—who developed the idea of a "vegetational continuum" and who would later have the University of Wisconsin's prairie restoration named for him—wrote, "The entire evidence of [our Plant Ecology Laboratory] study in Wisconsin can be taken as conclusive proof of Gleason's individualistic hypothesis of community organization."[150] Similarly, Robert Whittaker's "gradient analysis" supported Gleason's hypothesis.

Gleason's last two books—*Manual of Vascular Plants of the Northeastern United States and Adjacent Canada* and *The Natural Geography of Plants*[151]— were coauthored with Arthur Cronquist after Gleason turned eighty. Gleason's experience and perception show. *Natural Geography* demonstrates a lifetime of considering, in the face of fierce opposition, North American plants and their associations. Here are a few observations from *Natural Geography*:

> When the botanist William Baldwin visited the Middle West in 1819, he was surprised to see how every boat-landing along the Mississippi and Missouri Rivers was already populated by European weeds.[152]

> Bessey, many years ago, indicated that the trees of the eastern forests were migrating westward along the rivers of Nebraska.[153]

> The sharpest and clearest boundary in the country, so far as known to the author, is the one separating the Eastern Forest from the Prairies and Plains in Illinois and Iowa.[154]

> [There is extensive] protection of trees from fire by water on the west side.[155]

The Chicago School of Animal Ecology

William Morton Wheeler was born on March 19, 1865, in Milwaukee, Wisconsin, and lived there until he was nineteen.[156] About his youth, he would later write:

> Owing to my persistently bad behavior soon after I entered the public school my father transferred me to a German academy [that] had a deserved reputation for extreme severity of discipline. After completing the courses in the academy, I attended a German normal school which somehow had come to be appended to the institution.[157]

Wheeler's academy had a small museum, and in 1884, in anticipation of an exposition in Milwaukee, Henry A. Ward brought an assemblage of stuffed and skeletonized mammals, birds, and reptiles, as well as marine invertebrates, from his supply house. Ward persuaded city administrators to purchase the collection to combine with the academy's collection, and together they formed the foundation for the Milwaukee Public Museum.[158]

Wheeler worked in the academy's museum, and when Ward and his exhibits arrived, Wheeler helped the entrepreneur unpack and arrange them. Ward was impressed with the young man's knowledge and enthusiasm, and offered Wheeler a job back in Rochester, New York, which he took. Wheeler began in February 1884 and described Ward's business as "not so much a museum as a museum factory," where the "standards of workmanship were higher than in any of the museums that had grown up in various parts of the country."[159]

At Ward's, Wheeler was responsible for identifying and listing the birds and mammals being processed. As he gained experience, Wheeler was made foreman and worked to identify and arrange the marine invertebrate collections of shells, echinoderms, and sponges; he also prepared catalogues and price lists for publication. At about this time, Ward hired Carl Akeley as an apprentice taxidermist. Akeley and Wheeler became fast friends. Wheeler would late write of Akeley: "Of all the men I have known—and my profession has brought me in contact with a great many—he seems to me to have had the greatest range of innate ability. Although he later became an unusual sculptor, inventor, and explorer, he would probably have been equally successful in any other career."[160]

Wheeler returned to Milwaukee, where he began teaching high school. There, he was introduced to arachnology (the study of spiders) and the behavior of solitary and social wasps. Wheeler also came under the influence of C. O. Whitman, director of the Lake Laboratory.

Wheeler was made custodian of the Milwaukee Public Museum in September 1887. A year later, Akeley arrived at the museum, and the two young biologists shared an apartment in the city. A year after that, Whitman and others had made Wheeler a hard-core morphologist, and in October 1890 Whitman convinced Wheeler to take a position at Clark University, in Worcester, Massachusetts, where he would receive his PhD. Wheeler would also follow Whitman to Woods Hole during the summers of 1891 and 1892, where Whitman directed the Marine Biological Laboratory.

When Whitman received a call from the newly opened University of Chicago, Wheeler again went with him. To broaden his education, Wheeler spent his first academic year, 1893–94, in Europe, where he studied at the Naples Zoological Station, under Anton Dohrn. Wheeler returned and spent five years at the University of Chicago before moving to the University of Texas, where he stayed four years. In Texas he began working on ants.

In 1903 Wheeler accepted the curatorship of Invertebrate Zoology at the American Museum of Natural History, where he arranged the Hall of Invertebrate Life. He spent five productive years there, then, in 1908, accepted the professorship of economic entomology at the Bussey Institution of Harvard University. At Harvard, Wheeler occupied one lab in the Museum of Comparative Zoology, another in the Biological Laboratories. He spent more time in the Biological Laboratories than at the museum because he could smoke there.

Wheeler was quiet, modest, and unassuming. He was fluent in German, Latin, and Greek, and could understand each of the modern European languages. In 1912 he was elected to the National Academy of Sciences. Wheeler was an avid reader, "possibly the most widely read member" of the Harvard faculty. He remained there in various academic positions until his death in 1937. While at Harvard, Wheeler published over three hundred articles, mostly on the biology and social life of ants, paving the way for E. O. Wilson.

Victor Ernest Shelford was born September 22, 1877, in Chemung, New York. Chemung was sixteen miles from Elmira, where Mark Twain spent his summers.[161] Shelford went to grade school four miles away, at Oak Hill, and for the rest of his life wished he had had a better basic education in composition, grammar, and rhetoric.

By the time he was a teenager, Shelford had developed persistence, a quick and practical mind, an enthusiasm for doing rather than watching, and a firm belief in himself. He spent his summers during the early 1890s working as a busboy at Chautauqua Lake. There, he was exposed to the big ideas of educators, politicians, ministers, and other lecturers on the Chautauqua circuit.[162]

FIGURE 24. Field biologists at Barro Colorado Island in 1924. William Morton Wheeler is on the far right. Smithsonian Institution Archives, image #SIA-92-12929.

Shelford had a muddled education, trained to teach high school before he finished high school. He entered Cortland Normal and Training School at Cortland, New York, in early 1895 and completed his teaching requirement in 1898. Then, in 1898–99, Shelford took fifteen courses at Waverly High School, graduating three months short of his twenty-second birthday. That fall, he enrolled at West Virginia University.[163] His uncle, William Rumsey, converted his Cornell degree into a faculty position at WVU's Agricultural Experiment Station. There, working for Rumsey, Shelford prepared lantern slides and curated insects. He took nine courses in zoology, two in botany, three in chemistry, four in French, and two in German. The well-known comparative neurobiologist J. B. Johnston taught Shelford's zoology courses. Under Johnston, Shelford studied a diversity of animals and made formal reports on recently published literature.[164] It was at this time (ca. 1900–1901) that Shelford became interested in ecology. During his last year at West Virginia, he tutored zoology and applied to several universities, including Harvard and Syracuse, that accepted him. He settled on the University of Chicago, following former West Virginia president Jerome H. Raymond, who had taken a faculty position.[165]

At Chicago, Shelford took the embryology course taught by the famous Frank Lillie (who would, in 1930, help found the Woods Hole Oceanographic Institution). When Shelford was a senior, he came under the deep influence of zoologist, natural historian, statistician, and future geneticist Charles

FIGURE 25. Victor Shelford (white hat) with University of Chicago graduate student colleagues in 1906. Used with permission of the University of Illinois Library Archives, image #0007446.

Davenport. Davenport suggested that Shelford examine color variation in tiger beetles. He did and found colors were due not to pigments, but instead to absorption patterns in the structure and nature of surface films on the elytra (wing covers). Shelford received his BS in 1903 and spent the summer working on tiger beetles at Woods Hole and Cold Spring Harbor, on Long Island.[166]

Back at the University of Chicago for the fall semester of 1903, Shelford chose Cowles and Charles M. Child as co-advisors. Cowles knew Shelford was a good, determined, field biologist, while Shelford, in turn, described Cowles as a "first class man"—the teacher who influenced him the most.[167]

Following Cowles's approach, and aware that no one had yet tied an animal response to plant successional patterns, Shelford began examining tiger beetle distributions and found larval distributions—determined by species-specific female choice in egg-deposition sites—to be the determining factor.[168] Shelford's dissertation, only twenty-seven pages long, was based on a second, parallel study on tiger beetle life history and larval ecology.[169] As he was preparing his dissertation, Child proposed that Shelford be retained at Chicago.[170] He was. During the winter of 1907, Shelford and his wife took a belated honeymoon to Europe, organized around museum visits to inspect tiger beetle collections.[171] When he returned in April 1908, Shelford's charge was to build Chicago's animal ecology program.[172]

Shelford was influenced early by Europeans, especially Alfred Brehm, Karl Semper, and Karl Möbius. He was also affected by the quality and the scope of the work being done by his downstate colleague Stephen Forbes.[173] From these influences and his own observations, Shelford gradually developed a conceptual framework to compare against new data. For example, Shelford's "law of toleration" (which I learned in the late 1970s had filtered first through Warder Clyde Allee, then through his student [my mentor] Dick Bovbjerg) stated that a species is found where its limits of tolerance are not exceeded. Bovbjerg would always add the following precondition: "If it was able to disperse into that region." Shelford and his intellectual descendants assumed the same factors controlling local distributions scaled up to determine regional patterns and range-wide distributions.

Shelford expanded his work to examine fish distributions relative to the age of streams and to the age of wetlands.[174] As his thinking matured, Shelford began considering "response phenomena" and wrote: "The classification which ecologists are striving to build up will serve a purpose in behavior, physiology, and ecology, analogous in this respect to that served by the phylogenetic classification in morphological thought. It should however be flexible rather than rigid and true to fact rather than schemes." He proposed to fix characters: first, by measuring the reaction of an organism to important environmental factors under rigidly controlled conditions; second, by testing this reaction in the lab to a series of graded environmental conditions; and third, by testing this reaction in the field along environmental gradients.[175] Three-quarters of a century later, Bovbjerg was still following this protocol to understand the behaviors of crayfish and pond snails.

In 1909 Shelford was promoted to instructor and his teaching responsibilities expanded to include underclassmen, upperclassmen, and graduate students. A year later he took a five-week trip out west and did fieldwork in Nebraska, New Mexico, California, Nevada, Idaho, and Utah.[176]

Shelford's first PhD student was Warder Clyde Allee, who began working on aquatic isopods (pill bugs). Allee studied the movement patterns of these crustaceans in a water tank where he induced an oxygen gradient. Soon Shelford and Allee were studying the responses of a diversity of aquatic animals—including fishes, crayfishes, snails, clams, as well as insect larvae and adults—to gradients of various environmental factors.[177]

Shelford helped Cowles lead the European plant ecologists on tours through the Indiana Dunes during the 1913 IPE.[178] There, he met Arthur

FIGURE 26. Members of the University of Chicago's Department of Zoology in 1946. Warder Clyde Allee is seated in his wheelchair in the second row, far left. My mentor, Richard Bovbjerg, is in the back row, fourth from the left. University of Chicago Library, Special Collections Research Center, image apf1-0562.4.

Tansley and developed an association with Cowles's IPE co-organizer, Frederic Clements.

Despite his successes, Shelford faced an uncertain future at Chicago. He wrote to Charles Adams at the University of Illinois, "Things are going ahead here but rather roughly. I think I will get my promotion but am not certain." Half a year later, he again wrote Adams, this time not so positively: "I am supposedly on the slide here."[179] Shelford published his first book, *Animal Communities in Temperate America*, to wide acclaim, but it was not enough to save his appointment at Chicago, and Shelford was forced to leave. Forbes, then seventy years old and a patriarch at the Illinois Natural History Survey, could spot talent anywhere, much less dropped on his doorstep. He convinced his successor, Henry B. Ward (no relation to Henry A. Ward), to hire Shelford as an assistant professor, and Shelford began his career at Illinois in time for the 1914–15 academic year. Even a great university whiffs every once in a while, although Chicago would retain Allee.

Shelford was a progressive Democrat with firm opinions. His daughter noted, "He had to be persuaded fully and carefully before changing his mind about a matter he had decided on."[180] Even in middle age, he remained thin, fit, and energetic; his approach to life was reserved, even-tempered, dependable, hardworking, and optimistic. He was an early riser, loved strong coffee, and bicycled to campus. Shelford had a good sense of humor and deep self-awareness. He also suffered from chronic sinusitis and stress-induced indigestion.

Shelford kept an open door for his students. With them he was measured and thoughtful. He had a curious way of contracting his upper lip against his teeth, as if he were about to smile. Shelford liked Sousa marches and enjoyed hosting students at his home. He was handy, doing the household plumbing, wiring, and carpentry, as well as maintaining the family car. He preferred writing at home in the evening.[181]

Forbes convinced Shelford to investigate agricultural pests, which Shelford did, faithfully, although the process was long and tedious and interfered with his other research. From 1918 until 1930, Shelford spent every other summer at Washington State's Puget Sound Biological Station, on Friday Harbor, where he taught marine biology. Around the San Juan Islands, in the northern portion of Puget Sound, Shelford became interested in sunlight penetration in relation to water turbidity and the effect it might have on the zonation of seaweeds. Among his many colleagues were his former Chicago advisor, Charles Child, and Libbie Hyman. (At Puget Sound, Professor Child joined Ms. Hyman for a daily dip in the cold ocean.)[182]

By the end of the 1920s, Shelford had begun his collaboration with Clements on *Bio-Ecology*. Shelford stressed the similarity of the responses of animals and plants to the environment, urging biologists to learn how organisms actually live in nature and to use this knowledge as an organizing principle of ecology. Shelford and Clements were motivated in attempting to "correlate the field of plant and animal ecology by the common belief that it would tend to advance the science of ecology in general." They worked together twelve years. The summers that Shelford did not spend in Puget Sound, he worked with Clements in Colorado.[183] The most serious challenge to Clements and Shelford's partnership was the conflict between Clements's top-down philosophical predilections and Shelford's bottom-up sensibility.[184] Shelford was practical while Clements was dogmatic and at times defensive.[185] Their collaboration culminated in the 1939 book *Bio-Ecology*.[186] While this book has been largely forgotten, its contents are marvelously rich, providing a fact-based cross section of where the science of ecology sat immediately prior to the Second World War.

Most applauded *Bio-Ecology*'s attempt to unify plant and animal ecology, the depth of knowledge conveyed, and the effort. But not everybody was pleased. G. Evelyn Hutchinson criticized Clements and Shelford's failure to address biogeochemical and metabolic (nutrient flow) processes, while Charles Elton condemned the terminology and the philosophy behind the terminology (the "implication of certitude and finality"). Shelford responded by leaving town—spending the summer on the tundra of Hudson Bay—and planning a trip to Panama.[187]

Shelford, no doubt influenced by Cowles's Botany 36, used his institutional responsibilities to explore North America in his field course, Zoology S111. One student thought that Shelford "fervently wished that he could experience first-hand every kind of natural habitat in North America."[188] As his biographer wrote, "To see Shelford through the eyes of his students is the fairest way to judge him as a teacher, and his long field trips were perhaps his most effective teaching method."[189] Shelford took his first field trips to southern Illinois, then into Tennessee, on Illini Coach Company buses. In 1936 Shelford led his first transcontinental field trip. Zoology S111 students stayed "in waterproof, insect-proof tents with sewn-in canvas floors, and electric lights" and were provided with "a cot, mattress, two cotton and one woolen blankets." Their meals consisted of "a regular breakfast of fruit, hot or cold cereal, sweet rolls and butter, coffee, and eggs, ham or bacon." Lunch was "carried as fruit, sandwiches, etc. with milk, iced tea or lemonade—served in the field." Dinner was robust—"meat, potatoes, two vegetables, a salad, dessert and [beverage]." The class lasted seven weeks, and they traveled to the southwestern deserts, the Colorado Rockies, and the prairies of Iowa and Nebraska. During the halfway point in the trip, they stopped near Flagstaff, Arizona, at San Francisco Mountain, where C. Hart Merriam had developed his concept of life zones (see above, fig. 10, p. 50).[190]

For over a decade, Jake Weber drove the bus during Shelford's field trips. He had this to say about the professor:

> He was a quiet man. Said only what he had to. If he said be up at 6 A.M., he meant 6 A.M. In the morning he'd beat on the tent top with a heavy switch to get the students up—it was very effective. . . . Most everyone went to church on Sunday: Protestants to an available church, Catholics to theirs. Now and then Dr. Shelford would go with me to town—for mail delivery, or his half ale, half beer draught at a local tavern. Otherwise he'd often have a beer just after arrival back from the field. He really tested the students' physical stamina. He was in fine shape.[191]

Shelford's field garb was unvaried: "dark wool trousers, puttees above his field shoes, short dark coat, and felt hat."[192]

In the early 1930s, Shelford began visiting the Canadian tundra, where among other things he worked on lemmings. Between 1930 and 1932, he graduated four more PhD students, including S. Charles Kendeigh—who stayed at Illinois and a decade later became Eugene Odum's advisor—and Samuel Eddy, who famously explored freshwater zooplankton communities and would become advisor to my Antarctic colleague, Joe Eastman.

About Shelford, Kendeigh would later write:

> . . . his biotic community and biome concept, although but vaguely conceived in those days, has become today's ecosystem ecology. . . . We graduate students soon learned that to get the most out of a Shelford field trip one must keep within a few feet as he marched rather rapidly across the prairie or through the woods. All kinds of pearls of wisdom would drop out including appropriate sarcastic remarks about reductionist biologists, or that "Woods Hole establishment," which he viewed as anti-ecology. I think the reason he was not too well accepted by many of the more conventional biologists was that his holistic ideas were ahead of [their] time, but methods to deal with ecosystem level processes had not yet been developed. In other words, his concepts were great but field methods of the day were inadequate.[193]

Shelford did what he could with what he had. In his work and to his students, he emphasized long-term studies, quantitative methods, a combination of field and lab studies, the importance of life histories, and the value of elucidating simple behaviors and food web linkages.[194]

In 1940, while doing fieldwork on Barro Colorado Island in Panama, Shelford's wife, Mabel, aged sixty-four, contracted malaria, slipped into a coma, and died. Shelford's grief was deep, and it took him a long time to recover. He assumed the chair of the Zoology Department, responsible for its fourteen faculty and sixty graduate students. During the summer of 1942, Shelford took his S111 students down to Mexico. Henri Seibert, who was an emeritus faculty at Ohio University when I was a postdoc there, was on this trip. He once described Shelford as "a complex of personalities—one moment as charming as a veteran foreign diplomat, the next as fierce and stern as a marine drill sergeant."[195]

As he aged, Shelford preferred to work alone, his chief relaxations being ecological research first, beers or tequila second.[196] He retired in 1946, at the state-mandated age of sixty-eight. Shelford suffered a stroke in late 1961 and recovered. A year later he had a second. Again, he recovered but was visibly weaker. A year later he published his final book, an epic, *The Ecology of North America*.

Shelford's first doctoral student, Warder Clyde Allee, was born on June 5, 1885, outside of Bloomingdale, Indiana. His parents were Quakers, which paved the way for Allee to attend Earlham College, in Richmond, Indiana, where he

FIGURE 27. Warder Clyde Allee (*back row, second from left*) and his ecology class in 1923. Libbie Hyman is sitting in the front row, left side, with her back to the camera. Reproduced with permission of the University of Chicago Library, image apf1-00112.

played football.[197] After he graduated in 1908, Allee attended the University of Chicago. He worked with Shelford and received his PhD in 1912. After a series of one-year teaching appointments at the University of Illinois, Williams College, and the University of Oklahoma, in 1915 Allee returned to Chicago, to Lake Forest College, where John Merle Coulter had once been president. For the next six years, Allee spent his academic year teaching at Lake Forest and his summer teaching at the Marine Biological Laboratory at Woods Hole. In 1921 he returned to the University of Chicago as an assistant professor. After a stint as a dean, in 1928 Allee was promoted to professor (an order of rank reflecting better times, when faculty, not administrators, controlled a university's curriculum).

Allee's research career began modestly in 1911, studying isopods in forest ponds around Chicago. This work reflected both Shelford's and Cowles's influence. The summer facilities at Woods Hole allowed Allee to ask questions under the controlled conditions of a laboratory, and he became an experimental ecologist. He also learned about marine invertebrates.

As with so many of the men highlighted here, Allee's life was marred by personal tragedy. In 1923 his ten-year-old son was killed while walking to school. Needing a diversion, Allee and his wife, Marjorie, spent the winter of 1924 at the field station on Barro Colorado Island in Panama. There, he drove spikes into a giant canopy tree, climbed it, and began taking environmental data. Together, the Allees published *Jungle Island*, a popular account of these experiences.

Two years after Panama, Allee began a series of papers entitled *Animal Aggregations*, which resulted in the 1931 book with the same name. As Karl Schmidt detailed, the conclusions Allee drew included

> . . . the repeated and conclusive demonstration that there is an unconscious need for the presence of fellow individuals in many species among the lower animals, and in most of the higher; that there is, in effect, a deleterious effect for *under-crowding* [a phenomenon now called the Allee effect] as well as the more familiar one of *over-crowding*. . . . More important was the demonstration of the reality of an unconscious cooperation, which he referred to as proto-cooperation, in a wide variety of animal forms.[198]

In 1930 Allee began losing muscle control, and his neurologist found a tumor on his spinal cord. His surgeon operated and he recovered fully. Two years later, his symptoms recurred; again surgery, again recovery. In 1938 his symptoms returned, and this time his neurosurgeon removed every trace of tumor, which left Allee paraplegic. For the rest of his life, he used a wheelchair. Eventually, Allee resumed his full schedule of teaching, research, and committee work.[199] When lecturing, Schmidt noted, "Once in his chair . . . he was completely oblivious of his disability, and his audience became equally oblivious."[200]

Along with Shelford, Allee was criticized by the otherwise reasonable William Morton Wheeler as a "silo and saleratus belt" ecologist—midwestern field biologists concerned primarily with the physical rather than the social environment. This is ironic, because today, among rank-and-file biologists, Allee is known through his *Animal Aggregations* for pioneering studies of social behavior in invertebrates,[201] while Wheeler's work, centered on social insects, has been largely overshadowed by the brilliance of E. O. Wilson.

Reminiscent of Bessey's Sem. Bot. at Nebraska, the Chicago ecologists met regularly off-campus beginning in the summer of 1939. Every other Monday, University of Chicago faculty, Northwestern faculty, and curators at the Field Museum met to discuss scientific publications, their current research, and other activities.

For Allee, perhaps the biggest blow of all came in 1945, when Marjorie died. She was crucial to his life and to his success in life. She wrote popular books for girls, which supplemented their income and compensated for the financial challenges of Allee's surgeries and recuperations. After her death, Allee's daughters cared for him.[202]

Allee's belief saw him through. He was profoundly religious and lived his life within the framework of the Quakers, where he "represented the extreme of liberalism with the minimum of preoccupation with theology." In his essay "Where Angels Fear to Tread," he wrote:

> Religion has much to learn from science in objectivity, in willingness and courage to follow evidence fearlessly, and even in judging what constitutes valid evidence. . . . Science has much to learn from . . . a religion characterized by unselfish living and honest thinking.[203]

After being forced to retire from the University of Chicago at age sixty-five, in 1950 he assumed the chair of the Biology Department at the University of Florida, in Gainesville. Three years later, he married Ann Silver, an old family friend. They had two good years together before his chronic kidney infection flared up. Allee sank into a coma and died on March 18, 1955.[204]

One of the most famous early University of Chicago biologists was never on their faculty. Libbie Hyman was born in Des Moines, Iowa, on December 6, 1888. Her father owned a clothing store, sold out, moved the family to Sioux Falls, South Dakota, failed in businesses there, then moved the family to Fort Dodge, Iowa, where Libbie grew up.[205] She had a strong interest in nature—first plants, then butterflies and moths. She graduated from high school in 1905 and, after discovering she was too young to qualify for a teaching position, went back to high school to take advanced German, then took a factory job. Coming home from work one day, a friend mentioned that the University of Chicago was offering a year's tuition to students who had graduated from well-regarded high schools, including Fort Dodge High. During the fall of 1906, she enrolled at Chicago. About this, she wrote, "To the best of my recollection it had never occurred to me to go to college. I scarcely understood the purpose of college."[206]

At Chicago Hyman tried botany, but, she explained, "I somehow made an enemy of the laboratory assistant . . . who tried to have me flunked." Then she tried chemistry but realized, "I am not suited by temperament for quantitative work." She then tried zoology, where "I met with much encouragement and decided to make a career of zoology. I have never regretted that consideration." She worked with Charles M. Child, who she felt "was the outstanding member of the Zoology Department, but his original ideas and thinking antagonized other zoologists and long prevented the recognition he deserved." Hyman received her doctorate in 1915 and in 1919 wrote *A Laboratory Manual for Elementary Zoology*, published by the University of Chicago Press.

Ten years later, she wrote a second edition. In 1922 she wrote *A Laboratory Manual for Comparative Vertebrate Anatomy*; twenty years later, she compiled the second edition, retitled *Comparative Vertebrate Anatomy*. Afterward, she said, "I never liked vertebrate anatomy and since 1942 have abandoned all contact with the subject, refusing to consider making a third edition."[207]

Despite her distaste for some subjects of her books, their royalties made her financially independent. After her mother died in 1927, Hyman left Chicago and the expectations of her younger brothers that she continue as their housekeeper. She toured Europe for fifteen months, then settled in New York, near the American Museum of Natural History and its magnificent library, to devote her time to writing the definitive work on the invertebrates. Forty years and six volumes later, suffering from Parkinson's disease, she wrote, "I now retire from the field, satisfied that I have accomplished my original purpose—to stimulate the study of invertebrates."[208]

Hyman, through her interest in invertebrates, may have suggested to one of Allee's talented undergraduates, Edward Flanders Robb Ricketts, that the abundant starfish, worms, and jellyfish inhabiting the tide pools of Monterey, California, could support a "modestly thriving" trade.[209] Ed Ricketts, born on May 14, 1897, in Chicago, was "from birth, a child of intelligence and rare charm. . . . He began speaking very young and began using whole but simple sentences before he was a year old."[210] Allee was Ricketts's favorite teacher, and for the professional Ricketts, everything began with Allee. The impression of the professor on his student was reciprocated. Even twenty-nine years later, in an interview with Joel Hedgpeth,[211] Allee remembered Ricketts as "a member of a small group of 'Ishmaelites' who tended sometimes to be disturbing, but were always stimulating." After the fall quarter of 1922, Ricketts left the University of Chicago without formally withdrawing.[212]

Ricketts was one of the first biologists to bring the young science of ecology to the Pacific coast.[213] He acquired his ecological viewpoint from Allee. Allee had published his *Studies in Marine Ecology*,[214] which summarized observations and fieldwork at Woods Hole. Ricketts carefully studied this work and used it to guide his own observations on the Pacific coast; Ricketts considered Allee's conclusions "applicable everywhere."[215]

During their first years on the coast, Ed and his family lived in Carmel. After a few years, they moved to Pacific Grove, to a cottage on Fourth Street near Junipero Avenue. In 1930 John Steinbeck and his wife, Carol,[216] moved into the family cottage on Eleventh Street in Pacific Grove. Steinbeck heard about Ricketts, and apparently through Jack Calvin, a mutual friend who would later

FIGURE 28. Ed Ricketts collecting invertebrates in one of his beloved West Coast tide pools. Image used with permission of Nancy Ricketts.

become junior author on Ricketts's first book, arranged to meet Ed at Calvin's home. Shortly after, Ricketts hired Carol Steinbeck as a clerical assistant.

Ricketts and Steinbeck shared a deep interest in integrating humans and the environment into a unified whole. This interest was compounded when, in 1932, Joseph Campbell moved into a small guesthouse, nicknamed Canary, next door to the Pacific Grove home of Ricketts, his wife, Nan, and their three small children.[217]

Ricketts's lab, besides being a rare source of income for everyone during the depression (they called it the "bank"), became an improbable intellectual and cultural mecca. Artists, writers, painters, musicians, scientists, and both graduate students and faculty from Hopkins Marine Station dropped by. Parties sometimes lasted days. Ricketts, Steinbeck, and Campbell would spend hours talking about metaphysics, psychology, art, history, and literature, especially Goethe's *Faust*.

Once Ricketts established his business, he realized a layman's field guide to the seashore was needed. Because Ricketts knew these animals, his friends urged him to write one and volunteered their considerable talents. According to Joel Hedgpeth, Ricketts modeled his book after the 1874 classic by Addison Verrill and Sidney Smith on the invertebrates of Martha's Vineyard Sound, as well as Allee's papers on the intertidal invertebrates of Woods Hole.[218] Ricketts felt that while even the most intelligent layman would not know the scientific name or taxonomic classification of any particular organism, they would easily be able to identify its habitat. And from these habitat characteristics, using his book, the collector could then match the newfound specimen with an illustrated species. It was a novel approach that opened the world of the seashore to everyone.

Ricketts submitted the book as *A Natural History of Pacific Shore Invertebrates* to Stanford University Press in late 1931.[219] At some point during its consideration, someone—Hedgpeth suggests at Stanford Press—suggested calling the book *Between Pacific Tides*, and the name stuck. A respected Hopkins Marine Station scientist, W. K. Fisher, reviewed Ricketts's manuscript for Stanford University Press. His assessment was harsh. In a letter dated December 2, 1931, Fisher wrote, "I read the manuscript with somewhat mixed emotions. The facts are authentic so far as I could see and on the whole it was fairly well written, barring a certain vulgarity in places which doubtless can be eliminated by the Editor." He then aired his big complaint: "The method of taking up the animals from the standpoint of station and exposure on the seashore seems at first sight very logical but from the practical standpoint it seems to me not particularly happy. Both Professor [George] MacGinitie and I were quite frank with Mr. Ricketts on this score." Fisher ended his review, "In any event I think the manuscript should be read by a professional zoologist. It must be remembered that neither of the authors can be classified in this category, although Mr. Ricketts is a collector of considerable experience."[220]

It was a severe rebuke from two major figures of Pacific coast biology— Fisher was director of Stanford's Hopkins Marine Station; MacGinitie was director of Caltech's Kerckhoff Marine Laboratory. Although both scientists drank with Ricketts, they held his ecological notions in low regard, a feeling they probably had for the whole field of ecology. But Ricketts had an inner toughness and was not intimidated; he revised the manuscript, adding sections, such as one on tides as an environmental factor,[221] while keeping faithful to his overall ecological perspective.

Slowly, frustration turned to joy. On April 25, 1939, he wrote to a friend in typical Ed-speak:

What is tides, and why are they between Pacific? . . . Yes, I heard there was a book out. And little as you'd believe it, I was one of the last to know. With its customary delicacy, Stanford kept its midwifery hidden from me, until Dr. Light at UC who is already using it in his classes, wrote me a letter of congratulation. But then, when I wrote to Mr. Stanford himself suggesting the mother-father would like to see the bouncing child, they advised that Light had received unbound copy only, so it's alright.[222]

Between Pacific Tides is today considered a classic in marine biology, and with over 100,000 copies sold, it is one of the best-selling books in the history of Stanford University Press. It has become a model for many subsequent seashore books.

By late spring 1939, both Ricketts's *Between Pacific Tides* and Steinbeck's *Grapes of Wrath*[223] had been published. Steinbeck was getting a lot of unwanted publicity, especially from conservative politicians and groups such as the Associated Farmers. On October 16, 1939, Steinbeck wrote, "Now I am battered with uncertainties. That part of my life that made the 'Grapes' is over."[224] He thought the solution might lie with his friend Ed: "I don't quite know what the conception is. But I know it will be found in the tide pools and the microscope slide rather than in men."[225] Ricketts wrote:

> Jon [sp.] has been saying that his time of pure fiction is over, that he'd like next to portray the tide pools, that his next work will be factual. . . . He said that in February he and Crl [his wife, Carol] would go to Mexico. I said that if they'd wait until March I'd go along. . . . He said, "Fine," that maybe too we could get in a little collecting. [The trip] changed around more and more from the idea of a motortrip down to Mexico City, to the idea of, primarily, a Gulf of California collecting trip. (Jon said, "If you have an objective, like collecting specimens, it puts so much more direction onto a trip, makes it more interesting.") Then he said, "We'll do a book about it that'll more than pay the expenses of the trip." And as we considered it, we got more and more enthusiastic about the whole thing.[226]

Four days before they set sail, Steinbeck hired the seventy-six-foot *Western Flyer* owned and captained by Tony Berry. Ricketts kept a journal. Two weeks into the trip, Steinbeck wrote, "We've been working hard collecting, preserving, and making notes. No log. There hasn't been time. It takes about eight hours to preserve and label the things taken at the tide. We have thousands of specimens. And it will probably be several years before they are all described."[227]

The trip covered forty-one days and nearly four thousand miles of coastline. Ricketts's log includes notes from all aspects of their experience: scientific observations (sketches of the environment and of species collected, and measurements), impressions of the people they met, and philosophical ramblings.[228] The *Western Flyer* docked back in Monterey on April 20. The task of sorting and identifying the more than five hundred species collected fell to Ricketts, and soon after unpacking, Ricketts began sending out unknown specimens to specialists.

Ricketts and Steinbeck built their book, *Sea of Cortez*, throughout 1941, and it came together rapidly. Toni Jackson, Ricketts's "common-law wife," began typing the catalogue and bibliography sections; later, Steinbeck hired her to type his narrative.[229] Alberté Spratt created the illustrations; W. K. Fisher, who had previously skewered *Between Pacific Tides*, contributed color photographs.

On August 22, 1941, Ricketts wrote to Steinbeck:

> It seems gratifying to reflect on the fact that we, unsupported and unaided, seem to have taken more species, in greater number, and better preserved, than expeditions more pretentious and endowed, as we were not, with prestige, personnel, equipment and financial backing. As Toni says: "Two guys in a small boat, with enthusiasm and knowledge."
>
> It appears that our unpretentious trip may have achieved results comparable to those of far more elaborate expeditions, and certainly more unified and ordered in an architectural sense. It may well prove to be, considering its limitations, one of the most important expeditions of these times.[230]

Sea of Cortez was published by Viking on December 5, 1941,[231] but the events at Pearl Harbor two days later and the subsequent entry of the United States into the Second World War overshadowed the book's release. While Steinbeck wrote to Toby Street, "The reviews of *The Sea of Cortez* are extremely good and lively,"[232] the book got little attention and sales were low.[233] Ricketts wrote, "Royalties on the book have so far been nil, and again I am afraid I fathered a financial flop."[234]

The mature Ed Ricketts was a remarkable and complex man, and largely because of Steinbeck, he has become a legend, making it difficult to distinguish the quality of the man. Hedgpeth pointed out:

> Steinbeck's portrait of Ricketts as "Doc" in *Cannery Row* is the only full treatment of a marine biologist in English fiction. In the popular imagination, Ricketts has become Doc, a loveable character who lived just as he wanted to live, getting enough to drink, eat, listen to and go to bed with, and in the end, to read. In the minds of some students, all this is what you do when you are a marine biologist and the learning comes just as easy as the wine, women and song.[235]

Following the release of *Cannery Row* in 1945, *Time* magazine reporter Bob de Roos[236] and *Life* magazine photographer Peter Stackpole visited Cannery Row. In an interview in 1975, Stackpole recalled, "We ran into Ed Ricketts, who was a close friend [of Steinbeck's]. Ricketts was quite friendly with us and allowed us the run of his house. We photographed him at work and down in the basement and in his lab where he was doing various experiments with sharks and various fishes." About Ricketts's daily life, Stackpole wrote, "I remember he had music going all the time. He was very partial to Gregorian Chants."[237]

Ricketts let De Roos shadow him during a trip to the Great Tide Pool at the tip of the Monterey Peninsula. De Roos wrote: "Ed Ricketts has the best eyes I have ever seen at work. He would sneak up on a tide pool which I swore was absolutely empty of life and point out dozens of nearly invisible transparent creatures." Meeting Ricketts made a deep impression on De Roos. "Doc and I talked for hours but I cannot remember what we talked about. He was a magnetic man, very easy to like, very hard to forget. I have Steinbeck to thank for that meeting."[238]

The Wisconsin School of Limnology

Edward Asahel Birge was born on September 7, 1851, in Troy, New York. He received a bachelor's and a master of arts degree from Williams College, in nearby Williamstown, Massachusetts, where he began his work on zooplankton. "His early interest in the planktonic crustacean and the chance which brought him to the shores of Mendota combined to start him on an exploration of the world in which lake plankton live."[239] Birge began teaching at Wisconsin as an instructor in 1875,[240] where he was known as "Bugs."[241] He completed his PhD at Harvard in 1878. During 1880–81, he studied in Leipzig with Carl Ludwig, researching the central nervous system of frogs.[242]

On his return to Wisconsin, Birge became a one-man department of biology, offering courses in zoology, botany, human anatomy, physiology, and bacteriology. Later, when more faculty were added and a separate Department of Zoology established, he became its first chairman. (At this time, William Trelease, who later became the head of the Missouri Botanical Garden, was professor of botany.[243]) Birge became more involved in administration,[244] and in 1891 became dean of the College of Letters and Science; from 1900 to 1903, he served as acting president; and from 1918 to 1925, he served as president.[245]

Birge was a well-read, broad thinker, and he knew how to put his thoughts to use; he was witty and skilled in debate.[246] Birge "wished his students to secure an all-round training, to get the breadth of view that comes only from a broad survey of various fields of knowledge. The specialist in pursuit of his own particular line digs his canyon of activity deeper and deeper, narrowing his vision more and more, until he loses his perspective on the broader problems of life."[247] Birge followed his own advice. While his vocation was scientist and administrator, his avocation was literature, and his distraction was theology.[248] He epitomized the bottom-up construction integral to the original concept of "university."[249]

Through the establishment of the Wisconsin Geological and Natural History Survey in 1897, and in his role as its first director, Birge began a broad program of acquiring basic information on the location and shapes (morphometry) of the lakes of southeastern Wisconsin. To get this work done, he hired a full-time biologist, Chancey Juday.[250]

As Birge would later say:

> No one could have had limnology less in mind than I did when in 1894 I started to work out, by quantitative methods, the annual story of the microcrustacea of Lake Mendota. This statement is not an example of modesty, but a mere acknowledgment of necessity; for the best authority tells us that the word *limnology* did not appear in English until more than a year after our work began. . . . [T]he fish life of any small lake comes directly from these small crustaceans, which in turn live upon the microscopic plants floating with them in the lake waters. . . . As our crustacea-catching continued into mid-summer we found that our booty began to disappear from the lower waters of the lake. The process continued until the lake became divided into two widely different parts. There was an upper lake, about thirty feet thick, whose water was warm and was filled with abundant plant and animal life. Below this lay an abrupt transition to the lower and colder half of the lake, which was not only cold but also without living plants and animals. . . . As Mendota cooled in autumn its upper and active stratum gradually became thicker, and the lake reached its full activity at all depths in late October or early November; an activity limited only by the lower temperature of the water. This process is repeated every year. This story, which Mendota told me without asking for it, was the revelation that sent me into limnology . . . when Dr. Juday returned to his permanent position on the Survey; work on Lake Mendota could be prosecuted regularly and vigorously; and it could be extended to some one hundred and fifty other lakes, chiefly in southern Wisconsin.[251]

Chancey Juday was born near Millersburg, Indiana, on May 5, 1871. He received all three of his degrees from Indiana University: a BA in 1896, an MA in 1897, and an honorary law degree in 1933. As an undergraduate, Juday worked with the prominent ichthyologist Carl H. Eigenmann, who by coincidence had established a biological station on Turkey Lake (now known as Lake Wawasee) only a few miles from Juday's home. Eigenmann introduced Juday to zooplankton, and it stimulated his interest in zooplankton, phytoplankton, lake productivity, fish growth, and water chemistry. Juday's early papers described Turkey Lake, then Winona Lake, where Eigenmann's field station was moved. Juday also described the biomass and daily cycle of vertical movements of zooplankton in Lake Maxinkuckee.

Juday started his career by teaching high school science in Evansville from 1898 to 1900. He then began a series of one-year jobs—first as a biologist for the Wisconsin Geological and Natural History Survey, where he studied the daily migration of zooplankton in Lake Mendota and other southeastern Wisconsin lakes.[252] He then went to the University of Colorado to study fishes, where he was initially an assistant, then an acting professor of biology. He continued moving west, becoming an instructor in zoology at the University of California, where he studied the fishes of Lake Tahoe and the marine zooplankton in the waters around San Diego and La Jolla.[253]

Juday returned to Wisconsin in 1905, where he was again, thanks to Birge, appointed biologist for the Wisconsin Geological and Natural History Survey, a position he held until 1931. In 1908 he received a second appointment as lecturer in limnology in the University of Wisconsin's Department of Zoology. In 1931 Juday was promoted to professor of limnology, a position he held until his retirement in 1942. His work began when ecology in general and limnology in particular were in their formative stages; he helped to form both the Ecological Society of America and the American Society of Limnology and Oceanography.

Juday studied lakes globally. From October 1907 to June 1908, he visited limnologists and limnological laboratories in Europe. Two years later, he sampled lakes in Guatemala and El Salvador. The paper he wrote from these data represents one of the first studies in tropical limnology.[254]

Juday had no admiration for "arm-chair" biology. He was convinced by his own experiences that progress in the understanding of inland waters was made only through hard work, attention to detail, and large temporal and spatial data sets. His approach included flexibility of approach and a focus on fundamental problems. "Impetuous enthusiasm was not a part of his nature."[255]

The early efforts of Birge and Juday as a team were concentrated on the Madison lakes, especially Lake Mendota, and on other lakes of southeastern Wisconsin. Their discovery of seasonal distribution of plankton in lakes led them to investigations of thermal stratification and chemical changes. In particular, in the summer they noted that Lake Mendota set up a thermal layering (stratification), with the warmest water on top (epilimnion), the coolest waters on the bottom (hypolimnion), and an in-between narrow region where the temperature changed abruptly (metalimnion, or thermocline). They discovered that the epilimnion and metalimnion contained oxygenated water, and within these

FIGURE 29. Edward Asahel Birge and Chancey Juday dressed against type. Birge, the former University of Wisconsin president, dressed as if he was Juday, the field man, who is here formally attired. Used with permission of the Wisconsin Historical Society, image #131160.

regions zooplankton established a daily rhythm of movement, toward the surface at night, and at depths during the day (presumably to avoid fish predation). They also discovered that the hypolimnion was anoxic, and no zooplankton lived there. During the winter, the thermal layering reversed, with the coldest water (ice) on top, and the warmest water at the bottom. Each spring and fall the lake "turned over," with water the same temperature mixing throughout the column. This discovery was "the most outstanding single contribution of the Wisconsin School."[256]

After 1917, Birge and Juday shifted their intensive survey efforts away from the Madison region and began extensive surveys throughout the state and elsewhere. From 1921 to 1924, they carried out an intensive chemical and biological investigation of Green Lake, the deepest (236 feet) lake in the state and also the deepest lake in the United States (exclusive of the Great Lakes) between the Finger Lakes of New York and the mountain lakes in the West.

They also investigated the Finger Lakes and made the first limnological study of West Lake Okoboji in Iowa—today one of the best-known lakes in the world.

Birge and Juday also began studying northeastern and northwestern Wisconsin lakes. In June 1925, in cooperation with the State Forestry Headquarters, a summer field station was established on Trout Lake. Juday served as its director until 1942, when he retired. Their goal was to survey the chemical and biological properties of a large number of lakes to begin understanding the range of variation in these properties.

Thanks to Birge and Juday, "Limnology in Wisconsin is the story of the small lake—of the *lakelet*."[257] C. H. Mortimer wrote:

> To summarize their impact on limnology in a few words is difficult; but I believe [Birge and Juday] will be chiefly remembered because [they] laid bare the mechanics of stratification, and showed how the living processes of phytosynthesis, respiration, and decay combine to produce a concurrent stratification of the dissolved gases. The Wisconsin partners will further be remembered for their chemical analyses and crop estimates of plankton; and for the extensive survey of water chemistry and plankton in northeastern Wisconsin.[258]

Birge and Juday's legacy is a scientific program of limnology featuring a diverse array of specialists, including geologists, physicists, chemists, bacteriologists, algologists, and plant physiologists. They produced more than four hundred publications on descriptive limnology and limnological processes. Birge summarized, "What [we were] trying to do is lay the broad foundation . . . for future scientists in order that they may successfully carry on the work that [we] have undertaken."[259]

What It Meant

The second legacy of field biology is that it has formed, and always will form, the foundation of ecology. Field biologists give ecology its basis of facts—the physical, chemical, and biological factors that control where and when species occur. Field biologists have always been the arbiters in the conflict between ecological fact and theory—while theory organizes and makes sense of facts, theory must also be accountable to facts. There is no such thing as Huxley in reverse—a beautiful fact destroyed by a nasty, ugly little theory. These corrections of theory by fact can take a while, as the conflict between Clements and Cowles/Gleason illustrated. Paul Samuelson got it right when he said, "Funeral by funeral, theory advances."[260]

But just as ecology was gaining traction in North America, ecosystems were losing their signature animals. Bison were being decimated on the Great Plains, passenger pigeons were being shot out of eastern deciduous forest canopies, and the great rookeries of the coastal plain swamps were eradicated for the plume trade. It was time to begin conserving the animals and plants we wished to keep around.

The Wildlife Biologists

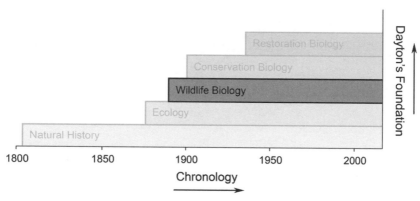

FIGURE C.

Americans ate passenger pigeons, wore the feathers of wading birds, and justified killing economically important bison through the narrowest definition of national interest. Could the extinctions of economically important species be justified in terms of short-term profits or political gain? Many Americans thought not; that it made more sense to preserve these animals, as well as the hundreds of fish, bird, and mammal species important either commercially or for sport. Out of this sentiment—due in large part to C. Hart Merriam's vision and political connections—arose the wildlife biologists.

The Scattered Become Gathered

While the U.S. Forest Service and U.S. National Park Service were formed in response to specific tasks—to oversee, respectively, our national forests and parks—the U.S. Fish and Wildlife Service did not have the luxury of pedigree. Instead, this agency arose from the need to understand the impacts of crop pests on the agricultural industry and was initially named the Division of Economic Ornithology and Mammalogy. Under Merriam's control and somewhat dishonestly, the division became a survey, documenting the distributions of birds and mammals across the continent. In 1905 the division was renamed the Bureau of Biological Survey to match its self-defined responsibilities. And once this mission drift began, it didn't stop. As federal protections for wildlife became law, Congress tasked the Biological Survey with their administration. For example, the Biological Survey became responsible for enforcing the Lacey Act of 1900, the 1913 Weeks-McLean Migratory Bird Law, the 1918 Migratory Bird Treaty Act, the 1931 Animal Damage Control Act, and the 1936 Fish and Wildlife Coordination Act. Then, reflecting the reality of these new responsibilities, in 1940 the Biological Survey was combined with the Bureau of Fisheries to form the U.S. Fish and Wildlife Service. No one covered more of this shifting administrative landscape than Edward W. Nelson.

Nelson was born in Amoskeag, New Hampshire, near the Merrimac River, on May 8, 1855.[1] When the Civil War began, his father enlisted and his mother moved to Baltimore to work in a hospital tending the wounded. Nelson and his younger brother were shipped off to his mother's parents' farm in the Adirondacks. Here, Nelson developed his love for nature during a trip to Loon Lake. "The exquisite beauty of the lake as the sun went down and night crept over the forest was beyond expression, and the wild, loud call of the Loon cleaving the twilight silence seemed to voice the spirit of this remote place."[2]

Toward the end of the Civil War, Nelson's father was killed and his mother moved to Chicago, where she opened a dressmaking shop. Nelson entered public school on the North Side, a few blocks west of the Lake Michigan beaches.[3] He developed his interest in birds first by shooting the common species in his neighborhood—his take included bluebirds, robins, brown thrashers, woodpeckers, nighthawks, rose-breasted grosbeaks, and scarlet tanagers—then by examining a friend's mounted birds and library, which included Thomas Nuttall's *A Popular Handbook of the Birds of the United States and Canada*, and Alexander Wilson's *American Ornithology*. Nelson tried to skin and mount a screech owl but made such a mess of it he gave up.[4]

Nelson's interest in bird specimens revived when he received a muzzle-loading shotgun and instruction in taxidermy. In 1872, while skinning some Wilson's phalaropes he had killed but delayed processing, he developed blood poisoning. The treatment of the day was recuperation in a dry, high-altitude climate out west. His high school principal used his connections to get Nelson added to Edward Drinker Cope and Samuel Garman's team, which joined Ferdinand Hayden's survey expedition at Fort Bridger, Wyoming. There, Nelson began collecting western birds. From Fort Bridger, Nelson went with Garman to Salt Lake City, then to San Francisco, shooting and preparing specimens along the way.[5] After Nelson returned to Chicago, he began corresponding with Joel Asaph Allen at Harvard. From Nelson's collection, Allen recognized a new species of sparrow, collected along the Calumet River in South Chicago, which he named *Ammodramus nelsoni*, after the budding naturalist.[6]

In 1874, while exploring the beaches of Lake Michigan, Nelson chanced upon Stephen Forbes and David Starr Jordan collecting fishes. For a short time after, Nelson became interested in fishes. He then developed a correspondence with Smithsonian ornithologist Robert Ridgway, and his interests returned to birds. Nelson spent the summer of 1875 with Ridgway's parents in Mount Carmel, Illinois, collecting along the floodplain of the Wabash River.[7]

A year later, another Nelson correspondent, Henry W. Henshaw, passed through Chicago on his way home after a summer spent on the Wheeler Survey, west of the 100th meridian. Nelson expressed his desire to become a naturalist, so Henshaw suggested Nelson go to DC and meet Spencer Baird. He did. Nelson also finally met Ridgway, and the two ornithologists hit it off. The following spring, Baird arranged for Nelson to acquire a position with the U.S. Army Signal Corps as a weather observer in Alaska. He stayed at St. Michael, southeast of Nome, for four years. There, Nelson became interested in the lives and customs of the Inuit. In 1887 he published his findings on their ethnology, as well as on the birds, mammals, fishes, and butterflies of western Alaska.[8]

When he returned to DC, Nelson contracted pneumonia, which developed into tuberculosis. He was not expected to live. But his Civil War nurse mother had seen worse. She packed him up, and they moved to the White Mountains of Arizona, where Nelson recovered, mostly. He had a residual heart condition for the rest of his long life.[9]

In 1890 Merriam appointed Nelson as special field agent on the Death Valley Expedition. Nelson joined Vernon Bailey in Keeler, California, on November 29, and the pair began collecting mammals, birds, amphibians, reptiles, and plants. Following the expedition, Nelson stayed in California to continue collecting. Two months later, Nelson began collecting in Mexico, where he worked for the next fourteen years.[10]

Few knew Nelson well. He usually worked in remote regions by himself or in small teams. He had strong convictions and in conversation was terse; Olaus Murie found him frustrating.[11] By the time Nelson became a DC bureaucrat, his personality was set. Among his few friends, however, he was

FIGURE 30. Edward W. Nelson on snowshoes and in decorated furs in Alaska, 1895. Photo used with permission of the Smithsonian Institution, image #SIU RU007364.

kind, warm, and deeply loyal. Teddy Roosevelt wrote that Nelson was "one of the keenest naturalists we have ever had and a man of singularly balanced development."[12] After a series of junior administration positions within the Biological Survey, in 1916 Nelson became chief, a position he held for the next eleven years. He used his influence to emphasize the catastrophic effects of swamp-busting on waterfowl, an issue that would preoccupy Nelson's FDR-appointed successor, Ding Darling, during the mid-1930s.[13]

Jay Norwood ("Ding") Darling was born October 21, 1876, in Norwood, Michigan, on Grand Traverse Bay. At that time it was a primitive, remote region, forty miles from the nearest railroad. His father was a schoolteacher who abruptly acted on his religious convictions and began preaching for Michigan Methodist. Darling's biographer brands him a "dreamer in a time and a place where hard work and long hours brought society's rewards, branded as lazy because he enjoyed reading and contemplation."[14] After Norwood, Darling's father served congregations in Cambria, Michigan; Elkhart, Indiana; and Sioux City, Iowa, where the family arrived in 1886.[15]

In Sioux City, young Ding received a "Pat and Mike" postcard from a friend who had gone to England to build Europe's first roller coaster.[16] Darling redrew the image and, from that point on, always had a sketch pad handy. His parents disapproved, believing "artists who drew pictures were classed with wicked playing cards, dancing and rum."[17] In those days this was a real concern. Sioux City was a rough, tough, wide-open frontier town, situated on the confluence of the Big Sioux and Missouri Rivers—the gateway to the northern Great Plains of Nebraska and the Dakotas. At the turn of the century, Sioux City was connected to Covington, Nebraska, an even tougher town, by a pontoon bridge without railings. Every morning entrepreneurs walked the bridge looking for bodies to fish from the Missouri; they were worth $1.00 each delivered to the police.[18]

As a boy, Darling explored the Loess Hills region of Iowa and, across the Big Sioux River, the prairies and bottomlands of southeastern South Dakota. Late in his life, Darling said these memories were his "pleasantest recollections."[19] Darling also claimed that during this time, he saw Missouri River "stern-wheelers loaded with bales of raw bison hides for the St. Louis market." If true (David L. Lendt unintentionally demonstrated through several passages that late in life Darling's memory was faulty[20]), these were certainly the last of the commercial bison harvested. By the time Darling's family reached Sioux City, in the mid-1880s, Teddy Roosevelt had given up ranching, and bison were far too scarce for hides to be shipped in memorable numbers.

During the summers, Darling mowed hay and did odd jobs for his uncle John back in Albion, Michigan. At first, he experienced clear streams filled with fish, migratory waterfowl, and songbirds. Later, he noted the farm was dying. Native trees had been cut, the stream was muddy, the well was dry, and the orchard was a tangle of dead limbs and tree stumps. Where there had once been eight inches of topsoil that produced sixty bushels of wheat an acre, the soil was eroded and unproductive. "This was my first conscious realization of what could happen to land, what could happen to clear running streams, what could happen to bird life and human life when the common laws of Mother Nature were disregarded." Sioux City was on its way to similar ecological deterioration, but when Darling was young, "life was rich and easy." By growing up where he did, when he did, Darling was a "witness to waste,"[21] and he carried the memories with him for the rest of his life.

In 1894, at age seventeen, Darling enrolled in nearby Yankton College, in South Dakota. He formed, captained, and played on the school's first football team,[22] then got kicked out of school for stealing the president's horse and buggy.[23] Darling next enrolled in Beloit College, Wisconsin—Stephen Forbes's alma mater—where "Ding" became his nickname.[24] He covered tuition and room and board costs by singing (bass) and playing mandolin. "I could sing in any religion you wanted, and I made the rounds of funerals every week." As a junior he was the leader of the glee club, manager of the track team, managing editor of the college paper, a member of the mandolin club and the Methodist Episcopal church choir, and the steady escort of Miss "Bit" Sumner—and he was flunking all of his courses except biology.[25] That year Beloit suspended him because of sporadic classroom attendance and poor scholarship, but mostly because of the humorous, spot-on campus newspaper sketches he drew of faculty and administrators. Darling spent the following summer barnstorming the Chautauqua circuit and his year of academic suspension as a reporter for the *Sioux City Tribune.*[26]

Darling's senior year was uneventful, and he graduated in 1900. Soon after he joined the *Sioux City Journal*—the *Tribune*'s competition—as a cub reporter. His first artwork appeared in the newspaper because a lawyer did not want his picture taken. Darling's first conservation cartoon, several years later, backed Roosevelt's campaign to establish the U.S. Forest Service. Darling later opined, "I think Teddy Roosevelt learned most of his forestry lessons from [Gifford] Pinchot, who later became Governor of Pennsylvania, and I am sorry to say [Darling's politics being the opposite of Pinchot's] not a very good one."[27]

As a Sioux City reporter, Darling traveled the prairie potholes of eastern South Dakota, where he claimed he heard of the young Paul Errington, who could shoot "like nobody's business."[28] Thirty years later, Errington would become the director of Iowa State College's Cooperative Wildlife Research Unit, Darling's innovative university/state/private partnership, designed to professionalize the field of wildlife biology (see below).[29]

Darling stayed with the *Sioux City Journal* for six productive years. On October 31, 1906, he married Genevieve Pendleton. While they were honeymooning, the *Des Moines Register and Leader* sent him a job offer.[30] They would double his salary provided he drew an editorial cartoon every day. At this time, the *Register and Leader* was a newspaper on the brink; it was in financial trouble and, as a result, salaries and morale were low, turnover was high.[31] Sensing that Ding was a game changer, chief editor Gardner Cowles gave Darling complete freedom to draw his cartoons, without approval from the editorial staff.[32] His first *Register* cartoon attacked the soft coal industry.[33]

Darling had been working at the *Register and Leader* for five years, establishing a national reputation, when, on November 2, 1911, he announced he was going to New York City to work at the *New York Globe and Commercial Advertiser*. But, being a product of a Great Plains upbringing, Darling never felt comfortable among the skyscrapers of New York City; he felt claustrophobic, hemmed in, closed off. He also suffered from asthma and from an increasingly painful and debilitated right arm. He stayed a little more than a year. The *Register and Leader* wanted him back, and he went,[34] although, because his cartoons were syndicated, he had to regularly travel back to New York.[35] At the time, he wrote about his return, "I still maintain the same privileges as I have always demanded—that I will express my own honest convictions in a cartoon without fear or favor and they can run the cartoon or leave it out as they choose. . . ."[36]

Darling was a big man for the time, six feet tall and 190 pounds, with a "dynamic, forceful walk."[37] He believed in self-reliance and was a strict and stern father, a perfectionist.[38] He became a pilot.[39] He smoked cigarettes incessantly, swore "eloquently and vigorously," collected fine books, and was an expert gardener. He disregarded newsroom deadlines[40] but got angry at anyone else's delays. He worked constantly and expected the same from others. He was a "vibrant, extroverted, outgoing, ebullient, strong personality"[41] and, as a result, was not a good team player.[42] If he liked you, though, you got his complete attention.[43] His personality brought a quality of motion, a sense of tension, to his cartoons; the impression that, if you took your eyes off it, the image would have shifted by the time you looked back.[44] According to his friend Rube Goldberg, Darling "developed the art of gentle ridicule better than anyone in the world. There is no defense against it."[45]

Darling won his first Pulitzer in 1923, only the second awarded for an editorial cartoon.[46] He won a second in 1942 for a cartoon he couldn't remember drawing.[47] That same year, Darling won the Roosevelt Medal for his conservation work and received a congratulatory letter from Pinchot (we can imagine what he thought of that).

FIGURE 31. Ding Darling sketching a cartoon. Permission to publish provided by the Ding Darling Wildlife Society, Doris Hardy, President.

Darling was a disciple of scientific method and felt that the "Laws of Nature" were the oldest legal code in the world.[48] As Lendt detailed:

> Darling . . . turned his avocation as an outdoorsman to the serious business of conservation of precious natural resources. . . . [He had been] distressed at the interference of politics in the life-and-death business of conserving limited resources. Officials in charge of saving irreplaceable water, air, and

land were hog-tied. Because of short-term political considerations, they could not institute what seemed to be only moderate and sensible programs for the long-term maintenance of America's natural bounty. Darling's love of nature and his distrust of politicians combined to create a fiery desire to see the politician's hands removed from the natural resources cookie jar.[49]

Darling was up to the task, described as "a man who can orate and he can snarl. He can bare his teeth and rear back on his haunches and let him drive at the malcontents and racketeers."[50] Darling saw the roots of wildlife issues in society: "All are taught in a manner equally oblivious of the fact that conservation of natural resources is a major factor in our national economy, standards of living, industry, international relations, and finally, peace and wars."[51]

Darling became instrumental in organizing the Iowa chapter of the Izaak Walton League of America, one of the most progressive and industrious. In the four years since its founding in 1922 by Will H. Dilg, the Ikes had attracted well over a hundred thousand members, several times the membership of any other American conservation group. William Temple Hornaday described Dilg as "a conservation John the Baptist, preaching in the wilderness." Dilg spoke to the midwestern hunter/fisherman who looked to older, simpler times and who feared that outdoor opportunities for youth were disappearing in an increasingly urban and pampered America.[52]

In 1931 Darling convinced the Iowa General Assembly to establish a nonpolitical state Fish and Game Commission. The original commission consisted of five members, including Darling.[53] In this role, Darling realized Iowa did not have enough scientifically trained personnel to offer a professional approach to wildlife research and management. He visited the president of Iowa State College and proposed the college participate in a cooperative program for research in wildlife conservation. The Darling proposal called for tripartite support among the college, the State Fish and Game Commission, and himself. He asked Iowa State to provide $3,000 per year for three years, which they did. Darling then secured an additional $3,000 per year from the commission and put up $3,000 per year himself. Paul Errington, recently graduated from the University of Wisconsin and a disciple of Aldo Leopold, became the leader of the nation's first Cooperative Wildlife Research Unit.

Iowans dominated the national political scene of the 1930s, and Darling knew them all. He was a Herbert Hoover man, but Hoover did about as much for ameliorating threats to wildlife as he did for improving the national economy.[54] When Franklin Delano Roosevelt was elected president in 1932, he appointed Iowan Henry A. Wallace, longtime friend of Darling's, secretary of agriculture.[55] This at a time when the Biological Survey was in the Department of Agriculture.

Before FDR, Darling was a political independent in his cartoons. But as he developed a deep distrust of Roosevelt and his socialist ideas, his drawings started attacking New Deal policies, which he felt destroyed private initiative and started the country down the slippery slope toward a welfare state.[56]

In 1934 Darling was appointed to the so-called Beck Committee "to devise a wildlife program that would dovetail with [FDR's] submarginal land elimination program." In 1933 Thomas H. Beck proposed that $12 million from employment relief funds be made available for a new wildlife restoration program and the program be run by someone independent of existing government agencies. FDR asked Wallace and Paul G. Redington, then chief of the Biological Survey, to evaluate the idea. Redington complied. The original committee consisted of Darling; the chair, Beck, editor of *Collier's Magazine*; and fellow Iowan John C. Merriam (no relation to C. Hart Merriam), a paleontologist, who by then was a regent at the Smithsonian and a strong wilderness advocate. Merriam could not serve, and so another Iowan, Aldo Leopold, was enlisted.[57]

As Darling later described:

> Out of that committee came the first enunciation of the essential program of restoring ducks' nesting grounds, it being recognized that the drainage program in the duck nesting areas had reduced the nesting areas to a very small percent of their former acreages. On the basis of that conclusion, we solicited advice from every state conservation officer or commission in the whole United States as to areas which might be subject to restoration, and laid out quite a program of possibilities. It was a hasty job and we knew at the time that much of the ground which we recommended for restoration was in agricultural production and probably too high priced for restoration to marshes and lakes.[58]

The Beck Committee meetings became contentious because its members (three men not used to compromise) could not agree on general policy. Beck, president of "More Game Birds" (which evolved into "Ducks Unlimited"), thought the way to restore duck numbers was to raise them artificially and release them into migration routes. In contrast, Leopold and Darling felt nature could raise waterfowl better than men could and advocated restoring the environment necessary to support them. These habitats included prairie pothole breeding grounds, upland feeding areas, migration rest stops, and wintering grounds. Darling eventually "took the whole bunch of [Beck

Committee] stuff home and wrote what I hoped would be a compromise." In his draft, Darling emphasized a program of restoration and refuges. But restoration was going to be expensive, and previous attempts at such programs had stalled. Frederic C. Walcott—U.S. senator from Connecticut, member of the Boone and Crockett Club, and a founder of the American Game Protective Association—led a campaign to create the Senate Special Committee on Conservation of Wildlife Resources and became its first chair. His counterpart in the House was A. Willis Robertson of Virginia. The solution to the funding problem became the creation of a Duck Stamp, which specifically targeted waterfowl hunters. (The less-desirable alternative plan, a 1% tax on sporting ammunition, would have more broadly taxed non-waterfowl hunters, including upland and big game hunters.[59])

In early 1934, Darling received a telephone call from FDR asking him to head the Biological Survey, at an annual salary of $8,000. Darling felt conflicted. He was the Rush Limbaugh of his day and was fiercely critical of Roosevelt's politics. Darling must have known Roosevelt was playing the game of keeping his friends close and his enemies closer. On the other hand, Darling felt he might have something to offer the New Deal team. He was financially independent, had a sense of and a feel for the problems he would be tackling, and could get things done. He knew this was a once-in-a-lifetime opportunity. Darling agreed to a temporary assignment on the conditions (similar to his newspaper stipulations) he be given full authority, independence, and funds to inaugurate the wildlife refuge program.[60]

There was resistance to Darling's appointment. Waldo L. McAtee—once one of Merriam's closest associates and head of the Division of Food Habits Research within the Biological Survey—submitted a lengthy report detailing Darling's incompetence.[61] McAtee was not a dispassionate observer. Following the Beck Committee's recommendations, McAtee wrote an alternative plan, with himself as chief, for the reorganization of the Biological Survey. He included a pointed criticism of any scheme submitted by a magazine editor (Beck) who got all his ideas from a capitalistic game hog (Leopold) or a cartoonist (Darling) who was a notorious critic of the New Deal administration.[62]

It's difficult to pinpoint how or why Waldo L. McAtee got on the wrong side of Ding Darling. McAtee was born on January 21, 1883, in Jalapa, Indiana. Soon after, his family moved to Marion, between Indianapolis and Fort Wayne, where he grew up fishing the Mississinewa River with his grandfather. When he was sixteen, he attended lectures given by the ornithologist Frank Chapman, which kindled his interest in birds. He attended Indiana University, and between his junior and senior years, he received a summer job as "biological

FIGURE 32. Waldo McAtee looking contemplative. Photo used with permission of the Washington Biologists Field Club.

expert" at the Biological Survey. After graduation, he joined the Biological Survey, where he remained until his retirement in 1947.[63] Two years after joining the Survey full-time, he received his master's degree; his thesis was on the relationship of agriculture to the presence and abundance of horned larks.

McAtee studied the food habits of ducks and geese, and made recommendations for plantings that would support them. In addition to his Survey responsibilities, he was acting custodian at the National Museum, where he worked on untangling the insect order Hemiptera (true bugs). He was the founding editor of the *Journal of Wildlife Management*. Over the course of his career, he published almost nine hundred papers. He continued his food habits studies of birds using stomach contents analysis. In 1932 he published a paper on insect camouflage as a predator deterrent. He was thorough; his data set comprised the stomach contents of eighty thousand birds.[64]

McAtee's social circles centered on the Washington Biologists' Field Club, at Plummers Island on the Potomac, not the trendier Cosmos Club. (Plummers Island would also become a favorite getaway spot for Howard Zahniser, coauthor of the 1964 Wilderness Act.) Possibly referring to his treatment by Darling, McAtee wrote:

> Perhaps to one like myself, for whom Nature heals all the troubles of the spirit, the greatest boon was the comparative quiet of [Plummers Island] and the adjacent mainlands. There was spent many a happy hour and pleasures experienced which, in recollection, outweigh contrasting experiences arising from the frailties of human nature.[65]

McAtee believed in the equality of all men. He did not treat subordinates as anything less than his equal, "for each man is superior is some way, knows more about something than any other. No man can be universal."[66] On the flip side, he expected to be treated as an equal. "If the alleged great wished to treat me as an equal, it was all right on that basis; if not, relations ended. This tended to alienate me from those in authority, for few of them remained unspoiled by it."[67]

Similar to Nelson, McAtee was shy and antisocial except around men who had similar interests and attitudes. He was a great friend of Francis Harper, who made his reputation speaking truth to power.[68] Consistent with Harper's attitude, McAtee detested small talk and frivolity. Once he made up his mind, it was made up; he was not prone to compromise. His eulogist noted there were times (perhaps best indicated by his assessment of the Beck Committee) when a "softened front would have been helpful to all concerned, and a loss to none."[69]

McAtee was viewed as a "naturalist of the old order." Indeed, he was a close friend and confidant of Paul Errington. He was energetic, competent, methodical, effective, independent, and courageous; he had a broad knowledge of nature, and, in the words of a friend, he "was the most honest man I have ever known."[70]

McAtee's opinions notwithstanding, Darling was appointed head of the Biological Survey on March 10, 1934, just four days after Roosevelt signed the new Duck Stamp Bill. As Lendt described, "Conservationists expressed guarded jubilation. They knew a man of stern stuff had been put in charge of a con-fused situation, and they knew Darling had the backbone required. But was Darling an administrator? Could he cut through the Washington red tape? Was anything really going to change for the improvement of conservation?"[71]

Darling sought advice from Ferdinand ("Gus") Silcox, chief of the Forest Service, who had offered sound counsel on Beck Committee issues. In re-sponse, Darling dismantled the Biological Survey's administration, reshuffled the staff (while ignoring civil service regulations regarding firings and reloca-tions), and hired new technicians. Darling felt that McAtee had been taking the credit while the excellent biologists under him were doing the work, so he re-placed him as head of Food Habits Research.[72] About this reorganization, Dar-ling commented, "Henry [Wallace] laughed when he saw what we were going to do with Food Habits Research [McAtee's position], and he didn't object."[73]

Clarence Cottam succeeded McAtee. Darling praised Cottam as "the most competent, efficient and courageous member of the present staff." Darling found another able man, J. Clark Salyer II, to head the development of the National Wildlife Refuge System. Darling called Salyer "my salvation." Aged thirty-two, Salyer became the first head of the federal refuge system. At the time, federal refuges consisted of a few isolated, unstaffed areas set aside to benefit various species of wildlife. Darling helped by sneaking through a $6 million federal allocation to purchase refuge lands.[74] By 1961, when Salyer stepped down, the federal refuge system had grown to 279 areas encompass-ing 29 million acres.[75]

When Darling took over the Biological Survey in 1934, it had a total of twenty-eight wardens in the United States and Alaska (at the time, a territory). In the face of no effective federal deterrent, market hunters were decimating waterfowl during both fall and spring migrations, in violation of the 1918 Mi-gratory Bird Treaty Act. Further, state officials were often hostile toward the federal government and the Biological Survey, and conspired to set their own seasons and bag limits for waterfowl, in violation of federal regulations.[76]

Darling responded to these threats by organizing secret strike teams—mobile squads of federal agents tasked to respond to trouble spots. Darling's teams were spread too thin to hit everywhere, and so they decided to make a statement. They conducted a surprise three-day raid resulting in forty-nine arrests. One suspect tried to shoot his way out (apprehending armed men, even hunters, is always a tricky business) and was critically wounded. The na-tion suddenly realized that this new Biological Survey was serious.[77]

Despite these actions, the Biological Survey got New Deal scraps. They were allocated some Civilian Conservation Corps (CCC) camps and Works

Progress Administration (WPA) crews and a "more or less friendly attitude on the part of the resettlement administration." Most of what the Survey received it stole "with the cooperation of some of the subordinate administrators," while the top-level executives "turned down every request we made for cash allocations."[78]

Following several years of drought that ushered in the Dust Bowl, by 1935 waterfowl numbers were at historic lows. In response, Darling enacted the tightest waterfowl hunting restrictions the country had ever seen. The waterfowl season was limited to thirty days, bag limits were capped at ten ducks and four geese, live decoys and baiting were outlawed, and the capacity of semiautomatic and pump shotguns was restricted to three shots (one in the chamber, two in the magazine).[79]

Darling's nascent Cooperative Wildlife Research Unit idea prospered at Iowa State under Errington. As chief of the Biological Survey, recognizing the need for professionally trained field biologists nationwide, Darling proposed similar wildlife units with federal backing at nine land-grant colleges. To fund the effort, he organized a national wildlife conference dinner and invited members of the sporting arms industry. Among the attendees were representatives of Remington Arms Company, as well as DuPont and Hercules Powder Company. Out of this meeting came several important initiatives, including the federal Cooperative Wildlife Research Unit Program, the National Wildlife Federation, the American Wildlife Institute, and the North American Wildlife Conference—perhaps the most productive three hours in the history of wildlife conservation.

Much to his surprise, Darling obtained an agreement from every American arms and ammunition maker to pay 10 percent of gross receipts into federal conservation channels.[80] Darling was struck by the response of the ammunition and firearms manufacturers. They "don't do much dictating, or at least, I never found it so. They gave me considerable financial support while I was in the Biological Survey and tied no strings whatsoever to their contributions."[81] By late 1935, Darling's twenty months as chief of the Biological Survey were coming to an end. His spontaneous, self-confident, swing-from-the-hip style had alienated people that the Survey needed.

As with so many of these men, there was deep personal tragedy. Darling's son, John, was as brilliant as his father. He had attended Princeton because it excelled in natural science and fulfilled his father's ambitions for him by becoming a physician. While driving from Rochester, Minnesota, to Des Moines in mid-January 1939, near New Hampton, Iowa, John's car slipped on snowy, wet Highway 65 and hit a bridge abutment. He was thrown from the car and fractured the back of his skull, which produced post-traumatic seizures; consequently, John became an invalid.[82]

Darling's own large and boisterous personality belied a history of chronic illness. In addition to his ulnar nerve problem, he suffered a heart condition and asthma, which had him hospitalized at the Mayo Clinic for most of a month in 1921.[83] In March 1925, he contracted peritonitis and was "at the brink of death." He remained debilitated for a year, incapable of drawing.[84] Soon after he recovered, he developed influenza with complications from pneumonia.[85] In 1936 Darling was again in poor health and decided to become a snowbird, staying in Florida to relieve his respiratory system from the effects of Iowa's winters.[86] He discovered the wonders of Sanibel and Captiva Islands, which Lendt suggested were his favorite places on Earth.[87] By May 1942, Darling wished to create a National Wildlife Refuge out of these mangrove forests.[88] In 1976 the U.S. Fish and Wildlife Service established a 5,200-acre refuge, named after Darling. At the same time, the northern half of the refuge was designated the J. N. Ding Darling Wilderness Area. Since then, the J. N. "Ding" Darling National Wildlife Refuge has been combined with Caloosahatchee, Island Bay, Matlacha, and Pine Island National Wildlife Refuges to form the J. N. "Ding" Darling National Wildlife Refuge Complex.

One of Darling's favorite Biological Survey biologists was Ira N. ("Gabe") Gabrielson, born on September 27, 1889, in Sioux Rapids, Iowa. Located between Storm Lake to the south and the Okoboji Lakes Region to the north, this was a region of abundant fish and game, an inspirational place for a future wildlife biologist.[89] After graduating high school, Gabrielson attended Morningside College, where he fell under the influence of the ornithologist T. E. Stephens, who was on the summer faculty of the University of Iowa's newly formed Iowa Lakeside Laboratory, in Okoboji.[90] Gabrielson attended Lakeside while an undergraduate. Following his graduation from Morningside, in 1912, Gabrielson taught biology at Marshalltown High School in central Iowa, and when a position opened three years later, he joined the Bureau of Biological Survey.

Gabrielson had been with the Survey about twenty years when Darling became director. Darling called for written reports from his biologists recommending habitats suitable for restoration. After they had submitted their choices, the men were summoned to Washington. As Darling recounted:

> When the group of about ten men from the northwestern United States came, among them was a very large pachyderm type of man with decided bucolic

FIGURE 33. An unpublished Ding Darling drawing, penned from his hospital bed on June 5, 1954, and sent to his Minnesota friend Dorothy Thompson. Cartoon reproduced here courtesy of James Anthony (Tony) Thompson and John Boardman, Dottie Thompson's grandsons.

attributes, who walked into my office one early morning in 1934. He didn't look like much but when his turn came to speak for his section of the country he immediately justified his position by a very well-stated analysis of Oregon, Washington and the immediate vicinity of his region and told, in detailed terms, the great needs for each specific area and what the possibilities for restoration might be. . . . [N]aturally he got immediate attention when it came to the allocation of funds. He set up his restoration program before anybody else had even got started and his choice of personnel among the engineers and construction men was a thing to delight the Chief of the then Biological Survey.[91]

The "pachyderm" was Gabrielson, who had joined the Biological Survey in 1915 as an assistant in economic ornithology. Darling first brought him to Washington as a consulting specialist, then as his handpicked successor. In 1940 Gabrielson became the first director of the newly formed U.S. Fish and Wildlife Service, where he got on the wrong side of the predator control issue (see below). Six years later, he retired from the federal government to become the president of the Wildlife Management Institute, where he reversed his position on predator control. Gabrielson joined ornithological expeditions to the Andes, the Amazon, Europe, the Mediterranean, South Pole, and Alaska. He wrote four books and coauthored six others, all of which were on birds and/or conservation. Gabrielson died of heart complications in 1977. Ding Darling would have been both shocked and pleased to know his pachyderm lived to be eighty-seven years old.[92]

Another Survey biologist, perhaps their most influential, was Herbert L. Stoddard, born in Rockford, Illinois, on February 24, 1889.[93] Stoddard and his mother moved to Florida when he was four, and on the way south first went east to stop by Chicago during the 1893 World's Fair, the event that would eventually spawn the institution—the Field Museum of Natural History—that would propel Stoddard to national attention among professional biologists.[94] Stoddard spent his youth outdoors in Florida, collecting and keeping animals, and these early experiences influenced his later decisions on career and study sites. His family returned to Rockford in 1900, when he was eleven, and two years later, they were destitute and he was on his own. At the time, Rockford was a blue-collar city, rough and industrial.[95] Seeking a profession, Stoddard decided to go in a direction different from the employment opportunities offered by his hometown. He was determined that his life's work would be taxidermy, and in this pursuit he was guided by two books, William Temple Hornaday's *Taxidermy and Zoölogical Collecting* and John Rowley's *The Art of Taxidermy*, plus more directed instruction through a correspondence course out of Nebraska. The local drugstore sold canaries, and he made scientific skins from birds that arrived dead after being shipped from suppliers or had died while being offered for sale.[96]

Stoddard's paternal grandparents were living in Prairie du Sac, in south-central Wisconsin south of the Baraboo Hills, and Stoddard went there to

FIGURE 34. Ira Gabrielson, director of the U.S. Fish and Wildlife Service, fishing along the Upper Mississippi River. Photo part of the collection housed in the USFWS Digital Library, accessed through the National Conservation Training Center. https://digitalmedia.fws.gov/cdm/singleitem/collection/natdiglib /id/3819/rec/2.

find work.[97] In the fall of 1906, Stoddard met Ed Ochsner, taxidermist, bee-keeper, fur buyer, and the man who would, thirty years later, introduce Aldo Leopold to his shack. Ochsner lived in Prairie du Sac and was admired in the region. In addition to his various successful business interests, he was one of the Midwest's best shots. He seemed to know everybody, including the Ring-ling Brothers, whose circus was based in Baraboo. Ochsner agreed to teach Stoddard taxidermy for twenty-five dollars on the promise he not compete with him locally. To encourage his young charge, Ochsner stressed the oppor-tunities at the natural history museums in Milwaukee and Chicago, where a person with enough skill in taxidermy could find a job and make a living.[98]

Late in February 1910, during a period of intense cold and deep snow, Ochsner and Stoddard went to Baraboo, to the Ringling Brothers Circus win-ter quarters. Henry L. Ward (son of Henry A. Ward, no relation to Henry B. Ward), director of the Milwaukee Public Museum, and George Shrosbree, its chief taxidermist, were there.[99] Shrosbree had prepared hundreds of hum-mingbirds and trogons for the Guatemala exhibit at the 1893 World's Fair.[100]

Stoddard used these connections to begin working as a taxidermist for the Milwaukee Public Museum.[101] While there, Stoddard received Robert Ridg-way's *The Birds of North and Middle America*, which Stoddard knew from Ochsner's shop in Prairie du Sac.[102] After a couple years, Ward's directorship of the Milwaukee Public Museum, inspired by his earlier corporate experi-ence at the supply house his father began, became more overbearing and con-trolling than Stoddard was willing to tolerate[103] (repeating the sentiment that George Kruck Cherrie felt toward the elder Ward). In late 1912, Stoddard left Milwaukee for the Field Museum, where he began working with the N. W. Harris Extension. He was responsible for preparing the mounted life-history groups of vertebrates—birds, mammals, reptiles, and amphibians—taken around by truck to children attending Chicago schools. To obtain material for his displays, Stoddard collected in and around Chicago, southeastern Wis-consin, and northwestern Indiana, especially at the Dunes.[104]

Stoddard stayed at the Field for seven years. He increasingly became more interested in field-based studies than museum work—in being outside than inside. Wilfred Osgood, formerly of the Biological Survey and both critic and admirer of C. Hart Merriam, was in charge of the Mammals Division in the Department of Zoology at the Field.[105] Through Osgood and others, Stoddard knew that the Survey was working across the continent to inventory verte-brates. He was particularly interested in the Survey's Division of Food Habits Research under W. L. McAtee exploring the feeding behavior of birds.[106] By this time, Edward W. Nelson was chief of the Biological Survey.[107] Nelson had little use for anybody who was not completely devoted to his work, and

FIGURE 35. Herb Stoddard banding quail during the early days of his fieldwork. Reproduced with permission of Tall Timbers Research Station and Land Conservancy.

by this criterion Stoddard was useful. During the fall of 1923, McAtee offered Stoddard a job as field leader, a salaried position in the Survey. He accepted and in March 1924 Stoddard started the Survey-sponsored Thomasville Quail project at three Florida plantations—Inwood, Sherwood, and Sunny Hill—which totaled 16,000 acres.[108] A key component, but by no means the only novel technique, of Stoddard's management program was to reintroduce fire into the native longleaf pine stands. Stoddard would later reminisce, "Owing in part to our twenty-year quail investigations, the ecological principles of the region are now understood in broad outline by most open-minded persons familiar with southeastern wildlife. But in the early years of our investigations it was considered heresy by most professional foresters to recommend burning in the pinelands."[109]

One of Stoddard's Survey colleagues, Olaus Johan Murie, was born on March 1, 1889, in Moorhead, Minnesota, the child of Norwegian immigrants. He developed his passion for animals and the outdoors growing up in the place where Teddy Roosevelt had first sensed the West.[110] Murie studied biology at Fargo College, then, when his zoology professor, Arthur M. Bean, moved to Pacific University in Forest Grove, Oregon, and offered Murie a scholarship, he transferred.[111] At Pacific, Murie studied zoology and wildlife biology, graduating in 1912. He began his career as an Oregon State conservation officer. In

1914 W. E. Clyde Todd, curator of birds at the Carnegie Museum of Natural History, hired Murie to collect birds and mammals, and to make study skins from these specimens. On his first expedition, to the Canadian north, Murie was accompanied by three Ojibwa guides.[112] Reflecting the growing standards of specimen documentation, Murie also sketched and photographed these animals and their habitats.

Murie's group traveled widely and became acclimated to subarctic conditions. After his field season, his guides went home, but Murie stayed to gain more experience. He collected additional specimens as he explored the ecology and culture of the far North. In 1917 Murie returned to Canada with Todd, businessman Alfred Marshall (who covered half the expedition's finances),[113] and his guides. The group followed the Ste. Marguerite River until they reached the Labrador Plateau, which they crossed to reach the Moisie River. They reached the Hamilton River and, after collecting at Ungava Bay, finally reached Fort Chimo (Kuujjuaq), in northern Quebec. Murie's expedition collected 1,862 specimens, representing 141 species of birds and 30 species of mammals.[114]

In 1920 Murie joined the Biological Survey as a wildlife biologist. A reindeer industry was developing in the Nome region, and Survey boss Edward W. Nelson, who had been a naturalist there in the late 1870s, was concerned.[115] Nelson did not want the inferior genetics of domesticated reindeer reducing the fitness of native caribou through crossbreeding. He felt the Survey needed to learn what areas caribou were using, so the reindeer industry could be expanded elsewhere. Murie was assigned the task.

Murie traveled by ship to the region around St. Michael. He was self-sufficient. In the winter, he used a dog sled; in the summer, he hiked cross-country, living on blueberries, ptarmigan, and what other meat he could scavenge for himself and his dogs. He was tireless, developing a well-deserved reputation among bush Alaskans as a wilderness traveler. Two years later, Murie hired his half-brother Adolph as his assistant.[116] As Adolph later wrote, "It was not only the promise of high adventure but also the anticipation of [his] companionship that thrilled me."[117]

In 1921 Olaus (the "slim, blonde young man, not handsome but with the freshest complexion and the bluest eyes") caught the attention of Margaret (Mardy) Thomas. She wondered, "How can he be a scientist and look so young?" Their courtship centered on outdoor activities. Olaus called in an owl.[118] Mardy taught him how to dance.[119] Olaus taught her how to fall backward onto moss beds without bending her knees; Mardy, to her great disappointment, could not teach Olaus to love the taste of beer (we do not hold this against him).[120]

FIGURE 36. Olaus Murie (dark shirt) cutting a rug with his Inuit friends. Photo used with permission of the Murie Center, Jackson Hole, Wyoming.

When Olaus set off with Adolph to study caribou on Mount McKinley (Denali; which became a national park in 1917), Mardy wrote him long letters. He replied with two in five months.[121] But somehow she knew. Mardy visited Olaus in the field, where they discovered "they were in love—no life for them except together."[122] Olaus's friend Jim Hagen said to Mardy, "You go ahead and marry that fellow; he's a fine one. The only thing is, I don't know how you're ever going to keep up with him. He's half caribou, you know."[123] The Muries were married in 1924. Afterward, she began accompanying him on field trips. One night Olaus was late getting back to their tent, where Mardy was alone, in winter, frustrated with him for being away for so long. When he returned safely, she was momentarily relieved, then got angry. He responded, "Oh no, you know I'm always careful. It just took so long to get the hide scraped so I could carry it back. Things take longer than you think, you know."[124]

Everywhere Murie traveled, he set out mousetraps.[125] As with Cherrie, in the field Murie carried a sixteen-gauge shotgun with a ".32 shot shell" barrel insert for songbird loads.[126] He also carried a .30-caliber Winchester Model 95 for caribou, bears, and wolves.[127] Mardy wrote about him killing and skinning a mink, making a sketch of another mink, and shooting a gull.[128] It was almost always past midnight when he finished putting up specimens and writing field notes.[129] Murie's labels included the date, location, scientific name, sex, total length, length of tail, and length of hind foot. His catalogue was a hard-covered black book that, at the time Mardy described it, had 1,940 entries.[130]

Later in his life, Murie "wore his packsack like a part of his anatomy, for he had to carry his notebook, color pencils for sketching, extra films, and numerous containers and papers for collecting all kinds of scat."[131]

When their son, Martin, was born, Survey chief Nelson wrote to the infant from the Cosmos Club:

> An old settler welcomes you as a newcomer to this mixed-up old world. With such good scouts as your father and mother to guide you on the trails you must follow, you will miss many of the rough places, but for your own sake I hope you will find enough of them to make life interesting. You have all my best wishes for a happy and prosperous journey. I hope you will be at least enough of a naturalist to appreciate the infinite variety of interest the world provides and to get from it health, happiness, and contentment. With congratulations on your arrival and all good wishes for your future I take pleasure in recording myself as one of your friends.[132]

Later that year, the Muries took baby Martin with them to the Arctic.[133]

Among his other Survey responsibilities, Murie was a fur warden, enforcing laws protecting animals against illegal hunting and trapping, and promoting caribou populations. One practice employed by the Biological Survey to promote game herds was predator poisoning. Murie, although initially not vocal, abhorred the practice.

> I have a theory that a certain amount of preying on caribou by wolves is beneficial to the herd, that the best animals survive and the vigor of the herd is maintained. Man's killing does not work in this natural way, as the best animals are shot and inferior animals left to breed. I think that good breeding's as important in game animals as it is in domestic stock. With our game, however we have been accustomed to reverse the process killing off the finest animals and removing the natural enemies which tend to keep down the unfit.[134]

Murie understood that hunting by humans counteracted Darwin's natural selection (so do roads, for that matter). He also understood that habitat loss was the biggest reason for declines in herds of large ungulates. The "caribou's greatest menace is not the wolf nor the hunter but man's economic development, principally the raising of reindeer." As we will see, later in his career Murie became much more outspoken.

Murie did his master's work at the University of Michigan and was granted his degree in 1927, the same year the Biological Survey assigned him to research the Jackson Hole elk herd.[135] He was among the first to discover that these elk historically migrated into the mountains and that the resulting crowding fostered disease.

Adolph Murie also worked for the Biological Survey. In 1939 he was assigned to study the ecology of the wolves in Mount McKinley National Park.[136] By this time, there was widespread concern over the increase in wolves in Alaska and what that meant to domestic reindeer, as well as to the big game herds. Pressure was building for wolf control.[137] As the younger Murie narrated, there was "a wise provision of long standing in the policy of the National Park Service that no disturbance of the fauna of any national park shall be made. Consequently, before anyone could embark on a program of wolf control, a number of questions had to be answered." His first assignment was to determine the relationship between wolves and Dall sheep,[138] named after William Dall, who had accompanied Kennicott on his last Alaskan expedition and who had been the Harriman Expedition's uncontested expert on Alaska.[139] Murie's goal was to learn when, where, and how Dall sheep died.[140] This is a strategy many field biologists will know and is the basis for wildlife management—first learn about death, then devise a plan to help animals avoid it. In 1939 Adolph Murie walked approximately 1,700 miles to find the answer.[141] It took him thirteen carefully argued pages and seven tables in his *Wolves of Mount McKinley* to offer that "the sheep and the wolves may now be in [ecological] equilibrium."[142]

Next, Adolph studied the relationship between wolves and Arctic foxes at McKinley. As his brother Olaus had concluded about the relationships between big carnivores and their prey, Adolph determined that Arctic fox populations had not been harmed by the presence of wolves and that both species inhabited the region in abundance.[143]

In 1937 Olaus Murie became an activist. He accepted a seat on the governing council of the recently created Wilderness Society. Murie lobbied successfully against the construction of large federal dams within Glacier National Park, Dinosaur National Monument, Rampart Dam on Alaska's Yukon River, and the Narrows Dam proposed for the mouth of Snake River Canyon. He also helped to enlarge existing national park boundaries and to create additional new units. His expert testimony on Olympic National Park helped to convince President Franklin D. Roosevelt to add the temperate rain forest of the Bogachiel River and Hoh Rain Forest in the Hoh River Valley, in the state of Washington.

Lobbying for a natural boundary for the elk of the Grand Teton area, in 1943 Murie helped to create Jackson Hole National Monument (it was later upgraded to national park status then incorporated into the Grand Teton National Park). The Jackson Hole National Monument victory was especially gratifying because it had hosted Murie's long-term elk study. Before it

FIGURE 37. Adolph Murie ascending Eagle Summit, in the White Mountains of central Alaska, in 1923. Photo used with permission of the Murie Center, Jackson Hole, Wyoming.

became a national park, Murie and others encouraged John D. Rockefeller Jr. to purchase the land and donate it to the federal government. Murie was unaware that Rockefeller intended to create "a wildlife display," so tourists could easily view wild animals without doing much more than getting out of their car. Murie intensely opposed this idea, believing by making nature too accessible, it would reduce its value. He believed instead that national parks should preserve primitive nature, a philosophy underpinning the Wilderness Society.

Once the Jackson Hole Monument was established, Murie developed its management plan. Despite loud protests from local sportsmen, Murie banned hunting within the park. Even when the state of Wyoming, in the case *State of Wyoming vs. Franke*, claimed the additional land held no archaeological, scientific, or scenic interest, Murie stood by the decision. He maintained that the park had biological significance by hosting a diversity of birds and mammals. The court ruled it could not interfere in the matter, a win for the conservationists.

As executive director of the Wilderness Society, Murie continued to advocate for national parks. Murie stressed the economic value of national

FIGURE 38. Olaus Murie concentrating on his watercolors while on the deck of the *Brown Bear*, in the Aleutian Islands, 1937. The *Brown Bear* was built in 1934 for the federal Alaska Game Commission and was later transferred to the U.S. Fish and Wildlife Service. Photo part of the collection housed in the USFWS Digital Library, accessed through the National Conservation Training Center. https://digitalmedia.fws .gov/cdm/singleitem/collection/natdiglib/id/7845/rec/8.

preservation sites because he knew they had public appeal. In the case of Jackson Hole National Monument, he emphasized how new tourism was contributing to Jackson's local economy. Murie believed those who wished to "seek the solitude of the primitive forest" should have the ability to do so and that a democratic society should protect this right.

The conclusions the Murie brothers drew based on predator studies were backed by Paul Errington's data on their prey. Errington was born on June 14, 1902, in the eastern South Dakota prairie and lakes region called the Coteau, on a farm ("The Old Johnson Place") his mother's parents had homesteaded, adjacent to the Tetonkaha chain of lakes.[144] There, growing up, he found "fish and turtle skeletons," "pieces of petrified wood," and flint arrowheads.[145] He also developed an unusual fondness for paper shotgun shells, both new and spent, which he liked to sniff.[146] When Errington was eight years old, he contracted polio. As he remembered, "My main problem was relearning to walk; and my most rapid gains in walking were linked with my development as a shooter at the beginning of my teens."[147] On Saturdays, Errington and his friends went on long hikes, stopping to shoot rabbits or roadside litter with their .22 rifles. He initially struggled to keep pace, but after a year or so rebuilt his strength and stamina. About his rifle shooting, he would write, "My outstanding asset . . . was keenness of sight; what I could see . . . I could generally hit if loads, gun, and gun sights were right and if I could steady myself and take my time."[148] He began collecting guns and learned how to shoot a shotgun. He would later say, "I saw enough expert shooting to be deceived by the apparent ease with which wing shooting was sometimes done. I thought all a hunter had to do was point a gun, pull the trigger, and then confidently expect the shot charge to do the rest."[149] Not so. Proving his scattergun talents were not innate, during his first duck hunt, at age fifteen, Errington used 125 shells to down six ducks.[150]

Errington noted the animals around him, especially their abundance:

> I liked to observe the wintering behavior of prairie chickens as they gathered by the thousands in the bottomlands. Their afternoon and evening flights were of mile-wide flocks. They roosted in slough grass; fed in cornfields; and, during blizzardy weather, took shelter in the lee of strawstacks, thickets, and ditch banks. They could be seen covering strawstacks, and I have walked around a strawstack to find them—not too wild—on all sides of me.[151]

Young Errington built his own canoe and read outdoor magazines and books by authors such as Ernest Thompson Seton. By age thirteen, he had decided to become an outdoorsman. "I was already thinking that maybe I should quit school and go up North as a wilderness trapper. However, I was not ready to announce my professional ambitions at home for fear they would not be well received. (My mother and stepfather wanted me to go to college and study to become a professor.)"[152] His parents did approve of his furbearer trapping.

Errington made money trapping minks, muskrats, weasels, and skunks, and by age fifteen, he was seriously considering a wilderness life. He familiarized himself with the lifestyle by building a camp, based on descriptions by

Seton (e.g., use white ash for firewood, without beavering) on a Lake Teton-kaha island. He spent a month trapping, learning how to stay warm in the land of the wind-chill factor,[153] and surviving on a diet of muskrat meat, carrots, onions, jams and jellies, and pancakes.[154]

Errington's experiences started him thinking about

> the adaptations of living things for living. . . . And I began to see vaguely that there were rules of order behind natural interrelationships. Predation was not a simple matter of a predator having an appetite for a given kind of prey and then going out and killing a victim at will. A given kind of animal did not live just anywhere it pleased. Some things that at first looked simple were turning out to be not simple at all as I learned more about them.[155]

Errington was deeply interested in learning the complicated interplay between minks and their muskrat prey. He was, first and foremost, fascinated with the behavior of muskrats: "Quite early in my experience I learned that wintering crises for muskrats were worth studying."[156] Errington was especially interested in "winter-wandering muskrats . . . thin, often with wounds from encounters with their own kind . . . feet and eyes frozen . . . dying or perhaps lie newly dead on the snow."[157] He could read the signs of predation in footprints, scattered fur, and blood spots.

He also followed minks, and he found one mink snow tunnel

> . . . packed with frozen frogs—most of them entire, some with leg or foot bitten off, others headless or half eaten or with only black egg masses remaining. When the thaws of winter exposed the snow tunnels, some of the frogs revived and slithered over the muddy drifts and into the water lying over the ice. (I never did see where the minks were getting all these frogs, but I learned from an Indian friend years later that his people sometimes gathered up to a bushel of frogs from certain wintering pools.)[158]

In 1925 Errington enrolled at South Dakota State College, in nearby Brookings. He found he could pay for most of his undergraduate education with the income he realized from pelts, and this endeavor, coupled with his hunting and camping trips, allowed him to continue spending much of his free time at Lake Tetonkaha and the surrounding region he knew so well. In turn, the faculty at South Dakota State used Errington's talents to build their museum collection. "Backed by a scientific collecting permit, I supplied the taxidermist with examples of what birds I could get—from kinglets to Canada geese."[159]

During this period, Errington's interest in writing grew. "I fancied myself an authority on minks and muskrats and was anxious to put my wisdom down on paper. I also thought, in those years before ever having heard of a rejection slip, that I could occupy myself profitably by writing fiction."[160] But first, as Sigurd Olson had discovered (below), he had to learn his audience as well as himself. "Mine was all he-man fiction, with nothing of love or romance in it. My stories were of trappers, northern wilderness, or [boxing] fights. My early literary taste ran to modifiers. . . . The North of my fiction was foreboding, baleful, menacing, merciless; it was a land of hardship, desolation, and immutable silences, yet of haunting, exquisite, ethereal beauty. . . . When I ultimately realized that my manuscripts were awful, I destroyed them with a greater feeling of accomplishment than their creation had given me."[161]

In 1929 Errington entered the PhD program at the University of Wisconsin, supported by a three-year fellowship jointly funded by the Sporting Arms and Ammunition Manufacturers' Institute and the Biological Survey. At Wisconsin, Errington met Aldo Leopold. Many people assume Errington was Leopold's student, but this was not true. In Leopold's obituary for the *Journal of Wildlife Management*, Errington wrote, "I was never formally his student. Informally, I moved in on him, his home, and his library for hours at a stretch, talking 'shop' or anything else. I wasn't a restful satellite and [we] sometimes argued in an evening until neither of us could sleep long after going to bed, but he was gracious toward me and patient with my ex-trapper's social deficiencies."[162]

This graciousness toward Errington was not only exhibited by Leopold, but also by his family. Before Errington's first date with Carolyn Storm, a childhood friend who had a law degree and was an accomplished musician, the young man—smitten, culturally depauperate, and perhaps thinking about lingering effects of his childhood polio—asked Nina Leopold, Aldo's oldest daughter, to teach him to dance.[163] It worked! Errington and Storm were married in 1934.[164]

In 1943 Errington and Leopold argued about how to interpret their data on quail population dynamics. This data set—to be compared with Stoddard's breakthrough studies at Thomasville—was collected south of Leopold's Shack, near Prairie du Sac. Leopold tended to be a top-down thinker. For him, theory often guided observation, and his studies were extensive. In contrast, Errington tended to be bottom-up, with observation generating theory—his studies were intensive. (In fact, both approaches are necessary to grasp the entire picture, as I'm sure both men realized.) Robert Kohler noted, "To Errington, Leopold seemed to have an imperfect grasp of the facts. To Leopold, Errington seemed to be drawing unwarranted conclusions from imperfect data." Leopold withdrew from the project, which Errington reluctantly published by himself. The men remained friends and continued to deeply admire each other's work. Leopold asked Errington for comments on an early draft of his first book, *Game Management*, where he mentioned Errington or his research on thirty-nine pages.[165]

With his PhD firmly in hand and a letter of recommendation by Leopold, Errington was hired in 1932 by Iowa State College to become the director of its Cooperative Wildlife Research Unit—Ding Darling's innovative partnership designed to populate and professionalize the field of wildlife biology.[166] Iowa State's Co-op had a rough start. The unit had a room in the Insectary Building, which housed Errington and his students; a reference collection of bones, hair, and feathers; samples such as owl pellets; and stacks of sampling equipment, rubber boots, and waders. They first persisted, then thrived, and with Errington setting the example, the Fish and Wildlife Cooperative idea caught on and—as had the agricultural experiment stations developed through the Hatch Act of 1887—became a federal initiative hosted by land-grant colleges.

Darling arranged for Errington's unit to organize a survey of the biological resources of Iowa. This project was funded by an appropriation of $35,000 covered through revenues generated by hunting and fishing license fees. These data then informed a comprehensive wildlife plan. Darling noted years later, "It was the first complete job of its kind in the United States and but for the untiring efforts of the Izaak Walton League and the peculiar susceptibilities of Governor Dan Turner to the evangelism this pioneer job of surveying and planning for a definitive program of conservation could not have been accomplished."[167]

Errington remained at Iowa State for his entire career, and as the years passed, he conducted an epic series of wetland research projects across the Upper Midwest, Great Plains, and Canadian Prairie provinces, as well as in the U.S. Southeast. He spent an estimated 32,000 hours studying Iowa marshes alone. Arlie Schorger suggested that Errington was "as acclimated to marshes as the muskrat itself."[168]

Errington's fieldwork, building on Stoddard's findings, helped reshape our understanding of how predators affect prey populations. Before Errington, fewer predators meant more game—that is, predators controlled prey populations. Errington discovered the opposite—that prey numbers could control predator numbers; that predators eliminated prey excess (down to what Errington called the threshold of security), as well as culling the weak and the old. Errington's perspective challenged the flawed understanding of predation that led to the federal programs—so opposed by Olaus and Adolph Murie—designed to exterminate the big western carnivores such as wolves, cougars, grizzlies, and coyotes.[169]

One of Aldo Leopold's most gifted graduate students, Hans Albert Hochbaum, was born on February 9, 1911, in Greeley, Colorado, and grew up among the mountains surrounding Boise, Idaho.[170] He attended Cornell, where he studied science, art, and ornithology, and received a BSc in zoology under

FIGURE 39. Paul Errington looking interrupted while writing field notes along a wetland margin. Photo used with permission of Fred Errington.

Arthur A. Allen (who founded Cornell's Laboratory of Ornithology). For three years, Hochbaum was a wildlife technician with the U.S. Park Service. In 1938, at age twenty-seven, he became the director of the Delta Waterfowl Research Station at Delta Marsh on Lake Manitoba.

Delta had been established in 1931 by James F. Bell, a wealthy Minneapolis industrialist who had founded General Mills.[171] Bell was an avid canvasback hunter and offered his hunting camp at the southern end of Lake Manitoba to the newly formed American Wildlife Institute (later the North American

Wildlife Foundation) as a research facility. Aldo Leopold chose Hochbaum to develop the facility and the program. Hochbaum loved the Delta Water-fowl Research Station, writing to Leopold after arriving, "I can't begin to tell you what a wonderful place this is. I would like to spend 10 years here."[172] As Leopold's student, Hochbaum worked and researched at Delta Marsh during the summers and spent his winters in Madison taking courses and consulting with Leopold. He received his MSc in wildlife management in 1941 and stayed on as the director of Delta until he took early retirement in 1970, to pursue his interests in art and writing.

Hochbaum's research at Delta—conducted in part with another Leopold student, Lyle Sowls, and others—identified territorial, homing, and nesting behaviors of prairie pothole breeding waterfowl. As well, they examined impacts of disease, predation, and hunting, including the impact of lead shot. About Delta's research, Frank Rohwer, Delta's chief scientist, commented, "It's important to remember they were starting from ground zero. . . . Everything they saw was new. Most of the research done to that point focused on the wintering grounds."[173]

Rohwer felt that Hochbaum "was a fabulous scientist . . . one of those really bright, gifted people. He could write well, was a great artist and a strong advocate for what he saw on the prairies."[174] He was a tireless advocate for waterfowl. Hochbaum's biggest contribution during his early years was the critical importance of small seasonal wetlands to breeding ducks. He also pushed for the creation of habitat surveys and for monitoring spring breeding populations. Hochbaum was practical and developed the method of sexing and aging ducks through cloacal examination.

Hochbaum was crusty and had strong opinions, which he frequently expressed to waterfowl managers. He once returned a student's manuscript with the comment "please reorganize; this is a dog's breakfast."[175] "He was entirely principled and was not one to be put off by authority," says Rohwer. Hochbaum felt that "research is the search for the truth, and management is the application of truth," and he had no tolerance for anyone who didn't play by those rules. "He didn't back down from anyone." Hochbaum's propensity to stand his ground when others were retreating manifested itself early and often. "I have strong personal convictions concerning the conduct of wildlife research and the application of its findings,"[176] Hochbaum wrote to Bell in 1942. He twice quit his position at Delta.

Hochbaum may be best known for his criticism of manuscript versions of Leopold's popular essays, which, when arranged, would become *A Sand County Almanac*. As Curt Meine detailed, Leopold and Hochbaum had, for some time, spoken informally about collaborating on a book of essays, with

Leopold providing the text and Hochbaum the drawings.[177] During the first half of 1944, they corresponded extensively. Hochbaum was a direct and honest critic, and Leopold responded quickly and openly to his occasionally personal suggestions. Both seemed to enjoy this literary give-and-take, which "would alter not only the flavor of Leopold's essays, but also the process of self-examination that went into them." Considering the essays as a whole, Hochbaum wrote: "In many of these you seem to follow one formula: you paint a beautiful picture of something that was—a bear, crane, or a parcel of wilderness—then in a word or an epilogue, you, sitting more or less aside as a sage, deplore the fact that brute man has spoiled the things you love. This is never tiresome, and it drives your point deep." Then Hochbaum expressed his biggest concern: "Still, you never drop a hint that you yourself have once despoiled or at least had a strong hand in it. In your writings of the day, you played a hand in influencing the policies, for your case against the wolf was as strong then as for wilderness now. I just read they killed the last lobo in Montana last year. I think you'll have to admit you've got at least a drop of its blood on your hands."[178] Leopold responded, although not initially and not without some pushback, by penning, on April 1, 1944, his classic essay "Thinking Like a Mountain."[179]

In the summer of 1944, Leopold began shopping his essays and Hochbaum's drawings, first to Macmillan, then to Knopf. Both publishers felt that Leopold lacked a unifying theme or principle. That fall, Leopold and Hochbaum discussed an almanac format for the book. Two years later, the project was progressing slowly, and Hochbaum withdrew because of his responsibilities at Delta.

During the mid-1950s, Hochbaum and Errington were featured in *Life* magazine as the top environmental scientists in North America. Hochbaum was honored with the Order of Canada in 1978, and two years before his death in 1988, he received the American Wildlife Federation Special Achievement Award. Other honors included a John Simon Guggenheim Fellowship (1961) and—my personal favorite—the Manitoba Golden Boy Award (1962) "for making Manitoba a better place to live."[180]

Hochbaum's major professor, Rand Aldo Leopold, was born on January 11, 1887, in Burlington, Iowa, to first cousins Carl and Clara Leopold.[181] Leopold was "a precocious student, interested in many things, and good at most everything he was interested in."[182] His interest in the natural world reflected his family's activities, and growing up along the Mississippi River and its waterfowl flyway gave him every opportunity to explore and, later, hunt.[183] In high school, Aldo was introduced to the "disciplined natural science that he would eventually make his life's work."[184] Leopold developed an interest in forestry, and at the time the big school of forestry in the country was at Yale University. In 1904 he shipped off to Lawrenceville Preparatory School in Lawrenceville, New Jersey, where he spent an academic year and a half laying the foundation for an Ivy League education, although he noted, "The instruction in English and History is much inferior to that of the [Burlington] High School."[185]

In September 1905, Leopold began his studies at Yale. He excelled. The rigor, formality, and status of the program suited him. After receiving his bachelor's degree, Leopold returned to Yale and in 1909 graduated with his master's in forestry.[186] He went to work for Gifford Pinchot in the U.S. Forest Service, assigned to the new Southwestern District comprised of the Arizona and New Mexico Territories.

On July 19, 1909, Leopold reported to the year-old Apache National Forest, just outside Springerville, in the Arizona Territory. His title was forest assistant, and he carried out reconnaissance and timber cruising. He had problems with his crew. As Curt Meine stated, "The problem was not simply that he was a greenhorn, but that he was confidently inflicting his greenness on the others."[187] It was about this time that the incident Hochbaum was referring to occurred. Leopold shot a female wolf and saw in her dying eyes the "fierce green fire" that would haunt him the rest of his life.[188]

Leopold had two other dramatic life-changing experiences during the two decades he spent with the U.S. Forest Service in the Southwest. The first occurred during the spring of 1911. While on temporary duty in Albuquerque, Leopold met Estella Bergere. Soon after, he took an assignment as deputy supervisor at the Carson National Forest, near Antonito, Colorado; quickly, he became supervisor. Aldo courted Estella long distance, and on October 9, 1912, they were married in Santa Fe. Meine wrote, "She sensitized him to an extreme degree. She inspired him in his thought, in his senses, in his work, and in his ambitions, and she would continue to do so for thirty-six years."[189]

Leopold's second transformational experience was less pleasant. In early April 1913, after settling a range dispute in the Jicarilla district, Leopold got caught in a spring storm that lasted two days and included hail and bouts of rain, sleet, and snow.[190] As he rode back, first his knees swelled, then later his face and limbs became swollen. He took a train to see physicians in Santa Fe and arrived barely alive. He had contracted a case of acute nephritis (Bright's disease).

His recovery took a year and a half. Six weeks after his diagnosis, he had regained enough strength to board a train with Estella and travel to Burlington, Iowa, where he convalesced at his parents' home, and she gave birth to their first son, Starker. At his parents' house, Leopold sat on the east porch resting, reading, and contemplating the view of the Mississippi River, far be-

low. He had always been a solitary thinker; here, his thoughts turned to conservation and back to Carson National Forest, and it wasn't long before his
ideas of conservation began to include game species. In February 1914, Leopold was allowed to return to the Southwest but not to Carson.

Resuming work, Leopold shifted emphasis from forestry to wildlife as-
signments and began developing a new program based on cooperative game
management that became a model for Forest Service activities nationwide. At
this time, Leopold approached game management based on principles of forestry management, which was essentially a quantitative assessment of a natural
resource. "Game could bring nearly as much income to the region as timber
or grazing uses of the forests, he calculated, if enough effort, intelligence, and
money were committed to develop the resource. . . . [T]he idea was not merely
to rear game and then release it to be shot, but to manipulate habitat so that, in
effect, the game raised itself."[191] A key feature of Leopold's new program was the
extermination of top predators, including wolves and mountain lions.

For his successes in the Southwest, in January 1917, Leopold received a letter of congratulations from Teddy Roosevelt. In July the same year, William
Temple Hornaday's Permanent Wild Life Protection Fund awarded Leopold
its Gold Medal. In his acceptance speech, Leopold stated the ideal was

> to restore to every citizen his inalienable right to know and love the wild
> things of his native land. We conceive of these wild things as an integral part
> of our national environment, and are striving to promote, restore, and develop
> them not as so many pounds of meat, nor as so many things to shoot at, but as
> a tremendous social asset, as a source of democratic and healthful recreation
> to the millions of today and the tens of millions of tomorrow.[192]

In the next few years, Leopold worked on a wilderness protection program and soil erosion problems. It was the idea of wilderness preservation
that initiated Leopold's departure from Pinchot's utilitarian philosophy.[193]
Leopold felt the condition of the forest should measure the effectiveness of
the management. Consistent with this idea, Leopold's major accomplishment
was to formulate the Regional Working Plan designating the Gila Wilderness
Area—the first wilderness area in the nation. Meine described: "Here was a
sign of cultural foresight, a willingness to let a wild place be. While serving
the self-interest of those, including and especially Leopold, who enjoyed the
experience of untamed country, it was, in the larger view of history, a quiet
act of national magnanimity. No European nation ever could, or ever would,
proclaim such a wilderness."[194]

In the spring of 1924, the director of the Forest Products Laboratory in
Madison, Wisconsin, was expected to resign, and in anticipation of an eventual

promotion, Leopold was offered the position of assistant director. In April he accepted "for reasons he never fully explained."[195] Leopold assumed responsibility for several important Forest Products Laboratory programs, including promoting closer cooperation among the national forests, reducing forest products industry waste, and encouraging the use of lower-grade tree species.[196] But this was not his kind of work. As the tedium of his position increased, Leopold looked elsewhere for release. He worked behind the scenes with the Izaak Walton League, one of the most effective conservation groups operating anywhere, for early protection of the Boundary Waters Area.[197] At about this time, the Leopold family took up archery, an interest that would eventually lead to the purchase of the Shack and its property.

In early 1928, four years after Leopold arrived in Madison, the director of the Forest Products Lab had still not retired, and it became clear to Leopold that "the longer he remained at the Forest Products Lab, the slimmer his chances for advancement seemed."[198] He took a month's leave in an attempt to finish his manuscript *Southwestern Game Fields* and to assess his future. In April he announced that he had "no intention to continue in my present place." And as word got out, suitors began appearing. Representatives of the Sporting Arms and Ammunitions Manufacturers' Institute approached Leopold with an offer to conduct a national survey of game conditions. Leopold agreed and found himself "entering a field which did not even exist."[199]

Central to the issue of game was the old-school approach, represented by the passage of game laws, versus the approach Leopold advocated, which was management through the control of environment, with hunting being but one factor. Within months, Leopold found "evidence of the most important trend of the times: the intensification of agriculture was eliminating food and cover plants required by upland game species."[200]

In September 1928, Leopold returned to Madison, and the University of Wisconsin offered him office space. Leopold hired a secretary, Vivian Horn, who noted, "I was astounded at the amount of data he could collect, and how steadily he could work assembling the data and turning out his reports after his return." In February and March 1929, Leopold delivered a series of lectures on game management through the University of Wisconsin, his first official connection with the institution that would ultimately play such an important role in his life.[201]

The result of his association with the Sporting Arms and Ammunitions Manufacturers' Institute, Leopold's *Report on a Game Survey of the North Central States*, was published and distributed in the spring of 1931.[202] Leopold's *Report* received positive reviews: "No one had ever packed so many facts about game and habitat into a single book."[203]

Leopold turned to his next project, the completion of a textbook on game management. He conceived of the book as a much-needed unifying treatise, detailing the history, theory, and practice of game management. He worked tirelessly, and for the first six months of 1931 did almost nothing but assemble the new manuscript. Leopold sought a publisher, but the Depression affected the industry and companies did not want to risk releasing books on a subject with no publication history. Finally, in December, Charles Scribner's Sons agreed to publish *Game Management*, provided Leopold make several edits to reduce production costs and contribute a $500 subvention. Leopold agreed and, as Meine detailed, "signed the contract on January 11, 1932, his forty-fifth birthday."[204]

In March 1932, Leopold's funding from the Sporting Arms and Ammunitions Manufacturers' Institute ran out. He had some investments, including stock in the Leopold Desk Company, and although the family was inconvenienced, they were not uncomfortable. Leopold continued to work on the manuscript for *Game Management*, and during the summer months he finished his final revisions.

With the publication of *Game Management*, Leopold invented the field of wildlife biology. In the now-famous last paragraph of his first book, Leopold wrote:

> In short, twenty centuries of "progress" have brought the average citizen a vote, a national anthem, a Ford, a bank account, and a high opinion of himself, but not the capacity to live in high density without befouling and denuding his environment, nor a conviction that such capacity, rather than such density, is the true test of whether he is civilized. The practice of game management may be one of the means of developing a culture which will meet this test.[205]

With *Game Management*, Leopold not only created an entirely new academic discipline; he also finagled the first-ever academic appointment in wildlife biology, at the University of Wisconsin, in Madison,[206] when, on June 26, 1933, the Wisconsin Alumni Research Foundation approved $8,000 per year for five years to support a game management program.[207] The *New York Times* "hailed it as the one and only 'wild game chair.' "[208]

Leopold followed up *Game Management* by putting his ideas into action. Through Stoddard's old mentor, Ed Ochsner, Leopold found eighty acres on the Wisconsin River, just north of Baraboo and east of the Wisconsin Dells. The place had a shack—an old chicken coop—and the property took on the name of the building. Leopold and his family began spending weekends there. At the Shack, Leopold did two things. First, he began putting into action the principles outlined in *Game Management*. He planted trees and restored prairie—managing habitat for wildlife. He also thought deeply—beyond science—about game management. With his predawn coffee pot and his notebooks, you can bet he was doing more than simply recording birdsongs. Many of these thoughts, strung together, became essays, and many of these essays became his second book, *A Sand County Almanac*.

In "A Daughter's Reflections," Nina Leopold wrote, "In an essay found among my father's works, he had written, 'there are two things that interest me: the relation of people to each other, and the relation of people to land.' As a place to put such ideas to work, my father bought a 'sand farm' in 1935 along the Wisconsin River, first worn out then abandoned by our bigger-and-better society." Nina asked, "This land had been 'lived on' and been 'destroyed' by its former owners. If you were selecting a piece of land to purchase, would this be what you would look for?"[209] It was. Leopold leased the place and visited again in February. Again Nina:

> Our introduction to Dad's farm came on a winter day in [1935][210] when he drove his family—his wife and five children and the dog—to have a look at the recently acquired farm. We drove the 50 miles in a February blizzard, shoveling out of snowdrifts and finally trudging the last quarter mile on foot through the snow. We arrived to find the only remaining building, an old chicken coop, complete with chicken perches and waist deep with frozen manure. As far as the eye could see was corn stubble, cockleburs, broken fences, and blowing sand and snow.[211]

Leopold visited the Shack for five days in late April (spring break—a trend that would continue as Leopold's academic calendar and the time for spring tree planting coincided). A week after this trip he wrote on page two of his Shack journal, "Bought place through [Stoddard's mentor] Ed Ochsner." The price was eight dollars per acre. Leopold returned to the property once in May, twice in June, and twice in July. During these early—1935 and 1936—trips he initiated activities that would characterize most of their trips to the Shack: starting plantings (food plots and trees), improving the livability of the Shack (installing first a clay floor and later a wooden floor, and nailing batten along the outside of the building), discovering breeding birds, and noting the phenology (seasonal timing) of the native plants (timing of emergence, flowering, seed set, etc.) and migrating birds.

Leopold and his children kept a Shack journal. On the unnumbered first page of the first notebook (four were eventually filled, three in his lifetime), Leopold typed:

> The Leopold Shack lies in Secs. 33 & 34, T. 13 N. R 6 E, Sauk County, Wisconsin. 80 acres were purchased in April 1935, and later added to. After 1940 the property comprised 123 acres.

FIGURE 40. Aldo Leopold recording field notes at the Shack. Used with permission of the Aldo Leopold Foundation, image #2004-0011.

The observations recorded in this journal pertain to this property, and often also the immediately surrounding region south of the Wisconsin River.

The observations and entries were made by Starker, Luna, Nina, and Carl Leopold, by myself, and at times by guests whose names are given. The map used in this volume was drawn by Carl Leopold in 1940.

Entries to the Shack journals were made in pencil. For each entry, visitors to the Shack were written in the top left corner, the date on the top right. Visitors almost always included members of the family, indicated by their initials. Graduate students also visited. For example, Fred and Fran Hamerstrom, who were working with Paul Errington at the time, came many times. Art Hawkins's first trip may have been March 19–20, 1937. Bob McCabe and Bill Hickey, the students who would eventually take over Leopold's position at Wisconsin, were often at the Shack. Leopold took his classes to the Shack on Saturday May 11, 1940, and on May 1, 1943. Colleagues and friends came, including Bill Vogt, Herb Stoddard, and the British ecologist Charles Elton. The journalist and incomparable outdoor writer Gordon MacQuarrie visited the Shack on October 12–13, 1940. Perhaps reflecting a conversation with MacQuarrie, Leopold wrote, "Why do rabbits eat pines so much around the shack and not elsewhere? In winter they get a lot in the woods, but the whites [pines] along the road and in the birchrow are not touched." Sometimes the group at the Shack was identified only as "AL et al." Leopold occasionally went by himself ("AL"), or with his dog ("AL and Flick"). The Shack journals are carefully indexed by subject, and the pre-index pages of each journal contain tables, figures, and notes comprising the first cut of the analysis of the data contained within.

On April 21, 1948, the day Leopold died, the first line of the Shack journal entry reads: "4/21 <u>Weather</u> clear, calm, cold, frost on grass, 36° at 600 AM. SW wind." The last entry, likely the last words he wrote, were "<u>Bloodroot in Shower Bed</u> Closed 6 AM."

The next Shack journal entry is three months later, from July 10–11, 1948. Estella, Estella Jr., Nina, and Luna were there. It begins:

> Spent Sat aft, & Sunday warm, partly cloudy; calm Saturday; S breeze Sunday work done: Cut most of lawn—very dry, no rain for at least one week.

A year later, on July 2, 1949, in blue pen, Leopold's friends Bill and Marjorie Vogt wrote on page 1 of the fourth Shack journal:

> A memorable day at the "Shack," rounding out impressions half complete during the years and bridging the gap of the last twelve years. A sad return, yet because life and spirit go on, with the pines, the deer, the pheasants and the butterfly weed, here are still joy and peace, and the knowledge that a great tree will keep on growing, from these deep-sunk roots!

On the same page, with the same blue pen, Durward L. Allen wrote:

> What Thoreau found at Walden Pond we have found here. We have taken it with us but others will find it undiminished. Never to be forgotten is the shack and the warm friendship it harbors.

What It Meant

As with ecology, fieldwork will always form the foundation of wildlife biology. Nelson, Stoddard, the Muries, Errington, and Hochbaum began by studying animals; Leopold started out as a forester. They worked by accumulating knowledge, setting aside lands, and promoting protective legislation such as the Migratory Bird Treaty Act of 1918 and game bird protections (setting seasons and limits and establishing the Duck Stamp). Wildlife biologists also began using techniques developed by ecologists to estimate population size and health, as well as to determine critical habitat characteristics.

Once wildlife biologists showed us through these techniques how to save valued species, they set the stage for the conservation biologists to try to save them all (because all should be valued). Leopold the forester/wildlife biologist then supplied us with the philosophical underpinnings of conservation/preservation efforts in the form of his land ethic. Further, because nearly all of these wildlife biologists were federal employees, they got embroiled in the conservation/preservation tension surrounding predator control and clearcut logging, especially when logging impacted wildlife species protected by the Endangered Species Act. Many of them struggled with these issues; others stayed true. I think of Nelson as the carrier, Stoddard as the pioneer, Errington as the rock, Leopold as the brain, and the Muries as the heart of early wildlife biology. And Darling made it happen.

The Conservation Biologists

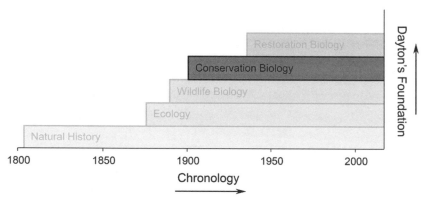

FIGURE D.

As it became clear that species extinctions at the hand of man should be avoided—that all species have inherent value and a right to exist—the question became how, in the words of Aldo Leopold, do we keep all these pieces? Answering this question has been, and will likely always be, a struggle.[1] The first "solution" came from Gifford Pinchot, who, through his Progressive perspective, took on the timber industry and its clear-cutting ways by designating national forests and establishing the United States Forest Service as well as a school at Yale to train foresters. But Pinchot's twin policy mantras of "wise use" and "the greatest good for the greatest number" also perpetuated ecologically destructive policies such as grazing in national forests. Further, Pinchot supported the damming of Hetch Hetchy Valley to provide water and hydroelectricity for the Bay Area, an action that mortified western preservationists, especially John Muir. The result was a half-century struggle to sort out land-use policies on America's federal lands, including national forests, national parks, national wildlife refuges, and wilderness areas. It is still a struggle. These were, and continue to be, bitter fights, producing civil wars among biologists and administrators mostly within, but also across and beyond, federal agencies.

Confrontations abated (but have not ended) only after agencies were forced to realize (by losing lawsuits) what the sweeping environmental protectionist legislation of the 1960s and 1970s meant for conservation policy. The big story has never been told, although its components have been well documented.

The Conservation/Preservation Tension

Gifford Pinchot was born on August 11, 1865, in Simsbury, Connecticut. Both the young man and his father were deeply influenced by George Perkins Marsh's *Man and Nature*, and as a result, his father suggested that young Gifford become a forester,[2] even though the profession did not then exist in the United States. Pinchot described the state of forestry in the late nineteenth century, before he entered and redefined the field:

> Outside the tropics, American forests were the richest and most productive on earth, and the best able to repay good management. But nobody had begun to manage any part of them with an eye to the future. On the contrary, the greatest, the swiftest, the most efficient, and the most appalling wave of forest destruction in human history was the swelling to its climax in the United States; and the American people were glad of it. Nobody knew how much timberland we had left, and hardly anybody cared. More than 99 per cent of our people regarded forest perpetuation, if they thought of it at all, as needless and even ridiculous. . . .[3]

He further explained:

> Thus the natural resources, with whose conservation and wise distribution and the whole future of the Nation was bound up, were passing under the control of men who developed and destroyed them with one and only object in mind—their own personal profit. And to all intents and purposes the Government of the United States did nothing about it.[4]

Pinchot's family was wealthy—in part from clear-cutting eastern forests—and after receiving his bachelor's degree from Yale in 1889, Pinchot went to Europe to learn forestry. He studied at the prestigious French L'Ecole nationale forestière, where he took three courses—silviculture, forest organization, and forest law.[5] In the field, he "ate lunches with the forest guards, wood choppers, and peasants, and saw the inside of practical forest questions through the eyes of the men who did the work with their hands."[6]

In Europe, trees are planted in rows, and Pinchot was taught to understand forests as a crop. In 1890 he brought this utilitarian view back to the United States. He first put his ideas into practice at George Vanderbilt's estate near Ashville, North Carolina. As designed by Frederick Law Olmsted, Biltmore was conceived as a model farm, a great arboretum, a large game preserve, and the nation's first managed forest.[7] In early 1892, Pinchot took charge. A year later, he opened an office in New York City to serve as a consulting forester, where he outlined never-realized plans for forestry schools at Columbia and the New York Botanical Garden. In addition, in collaboration with Henry Graves, Pinchot undertook studies of white pine and Adirondack spruce in the Northeast.

Based on his education, his well-regarded work at Biltmore, his studies on New England conifers, and his father's political connections, in early 1896 the National Academy of Sciences appointed Pinchot to the National Forest Commission, along with Alexander Agassiz and Charles S. Sargent of Harvard, William H. Brewer of Yale, U.S. Army General Henry L. Abbott, Arnold Hague, and ex officio member Wolcott Gibbs, who was appointed head of the National Academy of Sciences. Pinchot and Graves started west to the Northern Rockies in June, ahead of the other commissioners, to identify candidate forest reserves. When the full commission convened in Belton, Montana, on July 16, John Muir joined them.[8] The group then visited Pikes Peak before heading back East in October.[9]

On February 22, 1897, ten days before leaving office, President Grover Cleveland created these forest reserves exactly as the commission recommended. They added to an already existing 17,564,800 acres of forest reserves to make a national total of 38,844,640 acres.[10] At that time, all national forest assets were under control of the Department of the Interior, which did not employ a single forester.

There was, however, a federal Division of Forestry. In 1873 the American Association for the Advancement of Science sought to memorialize President Ulysses S. Grant by preserving forest. As Pinchot detailed:

> In 1876 Congress appropriated $2,000 for the salary of "a man of approved attainments and practically well acquainted with the methods of statistical inquiry," but not, you will observe, with Forestry. . . . In 1880 the Division of Forestry, in the Department of Agriculture, was created. Three years later Dr. N. H. Egleston, one of those failures in life whom the spoils system is constantly catapulting into responsible positions, replaced Dr. Hough. After three years of innocuous desuetude, Dr. Egleston in turn was replaced on March 15,

TABLE 1. The reserves identified by the National Forest commissioners and their size*

Site/State(s)	Areage
Bitterroot Forest Reserve in Montana and Idaho	4,147,200
Washington Forest Reserve in Washington	3,594,240
Lewis and Clark Forest Reserve in Montana	2,926,080
Olympic Forest Reserve in Washington	2,188,800
Flathead Forest Reserve in Montana	1,382,400
Mount Rainier Forest Reserve in Washington	1,267,200
Big Horn Reserve in Wyoming	1,198,080
Black Hills Reserve in South Dakota	967,680
Teton Forest Reserve in Wyoming	829,440
San Jacinto Forest Reserve in California	737,280
Uinta Forest Reserve in Utah	705,120
Stanislaus Forest Reserve in California	691,200
Priest River Forest Reserve in Idaho and Washington	645,120
Total Acreage	21,279,840

*Pinchot ([1947] 1998, 107–8).

1886, by Dr. Fernow, a graduate of the Prussian Forest School at Muenden. . . . Fernow saw forests as a state issue and the Division of Forestry as a bureau of information, "instituted to preach, not practice." Secretary [of Agriculture "Tama" Jim] Wilson was about to ask for Dr. Fernow's resignation . . . when Cornell opened its forest school and asked Fernow to take charge [as follows].[11]

With Bernard Fernow moving to Cornell, Pinchot was asked to head the Division of Forestry and, after some deliberation, he agreed. On Friday, July 1, 1898, Pinchot assumed the position that would come to define him (his two terms as governor of Pennsylvania notwithstanding).[12] But it would be almost two and a half years—on September 14, 1901, after William McKinley's assassination and Teddy Roosevelt's ascension to the presidency—before Pinchot got the administrative support he needed. And it would be another three and a half years after this before he got the legislative muscle he needed. On March 3, 1905, the Agricultural Appropriations Act gave all Forest Service men "authority to make arrests for the violation of laws and regulations relating to the forest reserves and national parks" and "in the enforcement of the laws of the States or Territories in the prevention and extinguishment of forest fires and the protection of fish and game."[13] This act also eliminated

FIGURE 41. Gifford Pinchot at his desk at the U.S. Forest Service office. Photo provided by Steve Dunsky and used with permission of the U.S. Forest Service.

legislation allowing for land swaps that favored corporations, and it renamed the Division of Forestry the United States Forest Service. Three months later, on May 31, 1905, Congress passed the National Forests' Transfer Act,[14] which shifted the administrative responsibility of all forest reserves from the Department of the Interior to the U.S. Forest Service. This act also dropped the term "reserves" from the name, making all these holdings national forests.

Pinchot regarded fire as an enemy of forests. Pondering the aftermath of a crown fire in the Priest Lake Forest Reserve near Blackfoot, Montana, he pronounced the scene "sickening."[15] Following the widespread destructive fires of 1910 (at least 3 million acres burned and eighty-five lives lost), he implemented the policy of fire control that would spawn the modern smokejumper and hotshot crews of today, as well as Smokey the Bear.

In 1894 Lieutenant George P. Ahern offered the first systematic instruction on forestry in America, at Montana State College, in Bozeman.[16] (Ahern would later become a severe critic of clear-cut forestry practices.[17]) Three years later, the Biltmore Forest School began offering one year of theoretical and practical forestry instruction. Cornell began its Forestry School the following year. At about the same time, Pinchot convinced his parents to donate

$300,000 to Yale to endow their Forestry School. In 1900, when the Yale school opened with seven students, Biltmore had nine, Cornell twenty-four. These early schools were shaky; Cornell closed in 1903, Biltmore in 1913.[18] In their place, other forestry schools emerged, usually at land-grant institutions and mostly farther west. In 1902 the "Forestry Course" was offered at Michigan State. In 1903 the Pennsylvania State Forest Academy was established in Mont Alto. In 1904 Iowa State, despite being located on tallgrass prairie, established its Forestry Department. In 1907 Harvard began its Forestry School. And in 1911, to replace Cornell, Syracuse opened the New York State College of Forestry (today the State University of New York hosts the College of Environmental Science and Forestry, located immediately south of the Syracuse campus).[19]

Pinchot was single-minded, and his views on the role of forests in the service of the United States never varied from utilitarianism or conservation in the sense of preservation for future use. He felt the nation's forests should be logged in a sustainable way and kept out of the hands of large companies and monopolies: "Better to help a poor man make a living for his family than help a rich man get richer still."[20] Pinchot's dictum, paraphrased from the "Father of the Scottish Enlightenment" Francis Hutcheson's "Inquiry Concerning Moral Good and Evil" (published in 1725),[21] became the motto for the U.S. Forest Service: "The greatest good for the greatest number."

John Muir was born in Dunbar, Scotland, on April 21, 1838, the third of eight children and first son born to Daniel Muir and Anne Gilrye.[22] During his childhood, he was "fond of everything that was wild." Young Muir loved "to wander in the fields to hear the birds sing, and along the seashore to gaze and wonder at the shells and crabs in the pools among the rocks when the tide was low; and best of all to watch the waves in awful storms thundering on the black headlands and craggy ruins of the old Dunbar Castle when the sea and sky, the waves and clouds, were mingled together as one."[23]

When Muir failed to learn his daily lessons, he was beaten: "The grand, simple, all-sufficing Scotch discovery [of the] close connection between the skin and the memory, and that irritating the skin excited the memory to any required degree. . . . [I]f we did not endure our school punishments and fighting pains without flinching and making faces, we were mocked . . . therefore we at length managed to keep our features in smooth repose while enduring pain that would try anybody but an American Indian."[24]

One night when he was eleven, Muir's father announced: "Bairns, you needna learn your lessons the night, for we're gan to America the morn!"[25] Sure enough, the next morning, Muir, his father, his brother, and his older sister Sarah took the train to Glasgow and made their way across the ocean and half of North America, to the edge of the frontier, where they created a farm on 160 acres of native woodlands and prairie, near Kingston, Wisconsin. Muir and his brother were immediately attracted to the unfamiliar nature. They considered how a woodpecker bored perfectly round holes and crowned kingbirds heroes for defending their nests against all threats, even hawks. One night his brother described a lightning bug meadow as being "covered with shaky fire-sparks." In reminiscence, Muir considered the display more impressive than the show put on by the "glow-worm lights in the foothills of the Himalayas, north of Calcutta."[26]

Creating fields from forests was difficult work, and Muir noted, "No pains were taken to diminish or in any way soften the natural hardships of this pioneer farm life; nor did any of the Europeans seem to know how to find reasonable ease and comfort if they would."[27] The epitome of this approach for Muir was his father making him chisel the family well, ninety feet deep, through sandstone. At eighty feet, Muir nearly succumbed to "choke-damp"[28] (a gaseous mixture typically composed of carbon dioxide, nitrogen, and water vapor).

Muir's father would permit only religious books in his house, so Muir would borrow "worldly books" from his neighbors and sneak them home. At age fifteen, he learned to appreciate the higher literature of Shakespeare and Milton. His father opposed late-night reading and, fully exasperated from constantly having to tell his son to go to bed, finally said, "If you *will* read, get up in the morning and read. You may get up as early as you like." Muir did, getting up at 1:00 a.m., giving himself "five, huge solid hours!" to read.[29]

At this time Muir began working on his inventions. He "made a fine-tooth saw suitable for my work out of strip steel that had formed part of an old-fashioned corset, that cut the hardest wood smoothly. I also made my own bradawls, punches, and a pair of compasses, out of wire and old files." He first made a self-setting sawmill, then thermometers, hygrometers, pyrometers, a barometer, clocks, a lamp-lighter, a fire-lighter, and an automatic horse-feeder. Muir's clocks were "regarded as a great wonder by the neighbors and even by my own all-Bible father." Muir's father encouraged him to take his inventions to the state fair, and when Muir wondered if people would value these things made of wood, his father replied, "Made of wood! Made of wood! What does it matter what they're made of when they are so out-and-out original. There's nothing else like them in the world. That is what will attract attention, and besides they're mighty handsome things anyway to come from the backwoods." Muir took his machines to Madison and displayed them in the Fine Arts Building on the University of Wisconsin's campus. "They seemed to attract more attention than anything else in the hall. I got lots of praise from the crowd and the newspaper reporters. The local press reports were copied into the Eastern papers."[30]

At the encouragement of a University of Wisconsin student, Muir made an appointment with Professor John Wheeling Stirling, acting president and dean of the faculty, where he asked permission to enroll. Stirling gave it, and Muir stayed four years, pursuing his interests in classical languages, botany, and geology, without ever declaring a major. Muir made two contacts at Wisconsin that would serve him well in California. James Davie Butler, professor of Greek and Latin, recognized Muir's love of books and introduced him to Emerson's writings. Later, when Muir was exploring the High Sierra, he brought along volume 1 of *The Prose Works of Ralph Waldo Emerson*.[31] Butler also encouraged Muir to keep journals. Muir's second Wisconsin influence was Ezra Slocum Carr, professor of geology and chemistry, who would leave Wisconsin to join the faculty at the University of California. Carr's wife, Jeanne, introduced Muir to West Coast scientists, urged him to write, and helped him get his first articles published.[32]

After his university experience, Muir landed a job improving machinery designed to make broom handles. After the factory burned down, Muir went to Indianapolis, in part because of its magnificent deciduous forests. There, he got work making carriage parts. He was so good, he was offered a partnership. One day a metal file slipped and injured his eye. Fearing blindness, Muir resolved never again to work indoors. Instead he dreamed of following Alexander von Humboldt and exploring the Amazon. Toward that goal, he began his famous walk (*A Thousand-Mile Walk to the Gulf*[33]). He made it as far south as Havana in January 1868, where he spent several weeks. He then continued on the schooner *Island Belle*, as she carried a cargo of oranges to New York. He left the city in early March on an emigrant ship and by the end of the month was in San Francisco. Muir walked south to Gilroy, where he may or may not have eaten garlic, then went east through the Pacheco Pass. As Edwin Way Teale described, "At the top of this eminence he had his first, never-to-be-forgotten sight of the great flower-filled Central Valley of California with the snow-clad Sierra Nevada mountains beyond."[34] Teale then argued that "the tragic suddenness of change" that turned wildflowers into mutton and "one sweet bee garden" into dusty ground awakened Muir's lifelong passion for preservation.[35]

Muir's first trip to the Sierras lasted about ten days. He spent the summer as a ranch hand, then returned in the fall, where he made arrangements to spend the following summer in the mountains herding sheep. Muir's journal of this time was published forty years later as *My First Summer in the Sierra*.[36]

The following summer, 1871, Emerson visited Yosemite and met Muir. Emerson was impressed with the height of the sequoias; Muir was disappointed at the briefness of the visit.[37]

Muir conducted geologic and botanic surveys to give his wanderings purpose. In 1871 he set to himself the task of exploring every peak and canyon in Yosemite, where he found glaciers.[38] Following up on this discovery, Muir began surveying glaciers, an activity that peaked in 1872 with his ascent of Mount Ritter.[39] In 1875 Muir shifted his interests and began surveying giant sequoias along their 200-mile front.

In 1879 Muir went to Alaska—the first of five trips. On this initial journey, he paddled north from Fort Wrangell with a group of natives. While the group rested, Muir climbed a ridge. When he got to the top, he became the first white person to see Glacier Bay; he also discovered what we now call Muir Glacier. Teale related that to the native paddlers, Muir was "The Great Ice Chief."[40]

Putting Muir in perspective, Teale pointed out, "The harsh life of his youth toughened his body to rawhide but, miraculously, it neither calloused or hardened his spirit. Wherever he went he saw nature with a poet's delight."[41]

On April 14, 1880, Muir married Louisa ("Louie") Wanda Strenzel and settled down in Martinez, California, on some land rented from his father-in-law. There, he raised Tokay grapes, Bartlett pears, and other fruit. By inventing special equipment for setting out orchard trees, taking care in packing his fruit, and developing new markets for his products, Muir was clearing $10,000 a year. He was the first to ship grapes from California to Hawai'i. About his business acumen, he would later state: "Why I'm richer than [Edward] Harriman. I have all the money I want and he hasn't. I might have become a millionaire but I chose instead to become a tramp. I have not yet in all my wanderings found a single person so free as myself."[42] Muir also wanted his landscapes to be free—Muir's wilderness was Muir's god—and he wanted it preserved.

Teddy Roosevelt credited Pinchot with being the "father of conservation,"[43] and many would argue that with his writings and by birthing the Sierra Club, Muir is the father of preservation (others would argue that either Emerson or Thoreau deserve this mantle). Pinchot and Muir met in person and initially got along well, but after Pinchot's reversal and subsequent advocacy of sheep grazing in national forests, Muir would have nothing to do with him. Any hope of reclaiming their relationship was destroyed in 1913 by the federal construction of the dam in Hetch Hetchy Valley, whose scenery had been every bit as spectacular as Yosemite's. At the 1913 legislative hearings on the proposal to dam Hetch Hetchy on the utilitarian grounds of providing

FIGURE 42. John Muir in Yosemite, 1907. National Park Service photo RL004398.

water and electricity to San Francisco, Pinchot testified: "If we had nothing else to consider then the delight of the few men and women who would yearly go to Hetch Hetchy Valley, then it should be left in its natural condition. But the considerations on the other side of the question, to my mind, are simply overwhelming. . . . I never understood Muir's position on Hetch Hetchy."[44]

From Muir's perspective, Pinchot's cozying up to San Francisco capitalists was perverse, and Pinchot's actions did not reflect the man Muir thought he knew. After Pinchot's dismissal from the Forest Service in 1910, Muir wrote, "I am sorry to see poor Pinchot running amuck after doing so much good hopeful work—from sound conservation going pell-mell to destruction on the wings of crazy inordinate ambition."[45]

Whither and How to Engage?

The first question for field biologists was whether this battle between conservation and preservation was one they should be fighting at all—is it possible to be both an objective scientist and an advocate for the systems you are
studying? The Ecological Society of America (ESA) at first thought yes. Two
years after the ESA was formed in 1915, members established the Committee
for the Preservation of Natural Conditions, with Victor Shelford as chair. The
committee encouraged members to send suggestions on valuable undisturbed
areas that were threatened by development and might be preserved. The result
was a list of around six hundred areas. Shelford's committee also initiated the
ESA's first lobbying effort, targeting government officials, legislators, and the
public on behalf of ecosystem and landscape preservation. The committee advocated for the formation of pristine areas in each national forest and national
park, was influential in securing preservation of Skokie Marsh, near Chicago,
and developed partnerships with the Okefenokee Society and Save the Redwoods League.[46] In 1920 the ESA voiced its opposition to altering natural water
flow by building dams and irrigation ditches. As well, the ESA disapproved of
the commercialization of national parks and monuments. Members encouraged other professional scientific societies to join this advocacy.

In 1926 Shelford's committee published its massive *Naturalist's Guide to
the Americas*, which provided a listing of areas that in their view should be
considered for preservation in Canada, the United States, Mexico, and Central America.[47] This ESA initiative arose at the same time Leopold was writing articles promoting wilderness preservation in national forests. While
the sentiments were similar, there were critical differences in emphasis and
priorities. For the ESA, "natural conditions" were defined as areas where
humans had not yet altered the ecological processes of succession and climax
communities.[48] A second difference was that the ESA wished to preserve as
many "representative types" of ecoregions as possible. Shelford listed twenty-
five specific "types of conditions" worthy of preservation, including tundra,
northern coniferous forest, deciduous forest, oak savanna, grassland, and a
variety of desert, forest, and wetland systems. The ESA's goal was to preserve
these areas as ecological reference sites.[49] The third difference was the representative ESA ecosystems being considered were small.[50] In short, unlike
Leopold, ESA ecologists were less interested in primitive experiences than
they were in un-impacted natural conditions.[51]

Twenty-five years later, the ESA did an about-face. Members decided
they must not advocate—that they should not be directly involved in saving
natural areas for research—and disbanded their preservation committees.[52]

Shelford was angry and in response organized a new society, the Ecologists' Union, to continue advocating and lobbying for habitat preservation. Eventually fences were mended, and five years later, in 1950, the Ecologists' Union and ESA and held a joint meeting on "Ecological Researches in Natural Areas." Afterward, at its annual meeting, the Ecologists' Union changed its name to The Nature Conservancy, which has since grown to become the largest environmental organization in the world.[53]

Once advocacy was decided, the second question for scientists and non-scientists alike became where to draw the line between utilitarian conservation and wilderness preservation. There were no easy answers and, as a result, this battle would play out throughout the first three-quarters of the twentieth century, concluding with the massive environmental legislation passed in the 1960s and early 1970s. Cornerstones of this legislation include the Clean Air Acts of 1963 and 1972,[54] the Wilderness Act of 1964,[55] the Clean Water Act of 1972,[56] and the Endangered Species Act of 1973.[57] In fact, this battle is still being played out.[58]

The conservation/preservation or Pinchot/Muir split turns out, in retrospect, to be not such a split at all,[59] but instead the right-side half of a human treatment-of-landscape continuum incorporating at least four classes of effect: obliterated, modified, conserved, or preserved. For example, mountaintop mining produces an obliterated landscape, swamp-busting produces a modified landscape (although wetlands are obliterated), selective logging produces a Pinchot-esque conserved landscape, while wilderness designation creates a preserved landscape.

One mistake commentators often make is assuming that a person's philosophy fits all situations.[60] While Muir had issues with his father's approach to discipline, he spent no time lamenting the old man's decision to clear-cut a portion of Wisconsin's old-growth eastern deciduous forest to settle the family farm.[61] What Muir and Pinchot were really debating was how to approach what they saw as the best-of-the-best landscapes, which in the early twentieth century were defined by scenic beauty. The tension these men created may have split the Progressive's environmental vision, but in the end, and looking at the big picture, the conflict was healthy because it either directly or indirectly birthed the four major types of federal landholdings designed to secure nature: national parks administered by the National Park Service, national forests and grasslands administered by the U.S. Forest Service, national wildlife refuges administered today by the U.S. Fish and Wildlife Service, and wilderness areas administered by all three agencies plus the Bureau of Land Management (BLM), which was created by the Taylor Grazing Act of 1934 to consolidate remaining unclaimed public domain under the Grazing Service (renamed the BLM in 1946).[62] Collectively, these areas encompass the con-servationist and preservationist visions of both Pinchot and Muir, offering something for everyone. How and why these areas formed is almost as inter-esting as the nature you find while exploring them.

America's Best Idea, Expanded

National parks stemmed from the idea for public city parks—"nature mu-seums"—for common welfare, which arose simultaneously in Europe and North America in the 1830s. A decade or two later, personalities as different as Thoreau and the landscape architect Andrew Jackson Downing were en-couraging municipalities to set aside tracts of green space.[63] Following a false start with Yosemite, the idea was first applied in 1872 with the designation of Yellowstone as a national park. It would take another forty-four years for the National Park Service to be born.

On August 25, 1916, President Woodrow Wilson signed into law the Na-tional Park Service Organic Act (simply, the Organic Act; Title 16)—and with it established the National Park Service. The Park Service was placed in the De-partment of the Interior, and its first director, Stephen Mather, was charged with supervising and maintaining all designated national parks, battlefields, historic places, and monuments. By 1916 there were already six holdings: Yellowstone (1872), Sequoia (1890), Yosemite (1890), Crater Lake (1902), Wind Cave (1903), and Mesa Verde (1906). A quick glance at figure 43 shows that since then, na-tional parks have been added at a regular pace, with new parks added one or two at a time, except for 1980, when nine new parks were designated. Today, in addition to traditional national parks, National Park Service responsibilities in-clude national seashores, national lakeshores, national rivers, national wild and scenic rivers and riverways, national trails systems, and national heritage sites.

In contrast to national parks, national forests arose in response to the tim-ber industry's aggressive logging practices. As described above, Gifford Pin-chot, John Muir, and many others were instrumental in establishing national forest reserves, which became our national forests. The Forest Reserve Act of 1891 gave the president of the United States the power to set aside forest reserves on federal land. The first national forests, established the same year, included the Yellowstone Park Timber Land Reserve in Wyoming, Shoshone in Wyoming, and White River in Colorado. The following year, Mount Hood in Oregon, Pike in Colorado, and Santa Fe in New Mexico were established. The 1897 Organic Act supplemented the 1891 legislation by protecting forests, securing water supplies, and supplying timber. With the Transfer Act of 1905,

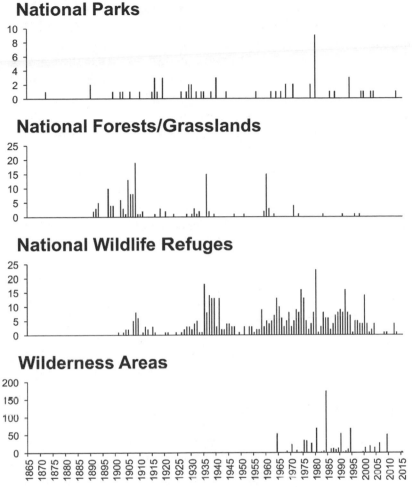

FIGURE 43. Chronology of additions of major federal properties, by type. Note that vertical axes for National Parks, Forests/Grasslands, and Wildlife Refuges are roughly equal in scale, while the y-axis of Wilderness Areas is increased by about a factor of ten.

forest reserves—previously Interior Department holdings—became the responsibility of the U.S. Forest Service in the Department of Agriculture.

Teddy Roosevelt saw conservation and preservation as two sides of the same coin. Using the power of the Antiquities Act,[64] which he signed into law on June 8, 1906, Roosevelt created national monuments from existing public lands to protect significant natural features, including national forests.[65] In 1911 Congress passed the Weeks Act, authorizing additions to the National Forest System in the East.

In 1937 the Bankhead-Jones Farm Tenant Act authorized the establishment of national grasslands, the herbaceous equivalent of national forests. Today, there are twenty national grasslands, most located on the Great Plains. National forests and national grasslands are administered identically, but, as Norman Maclean said about these habitats, "Different ways to live and die."[66] The Forest Service manages a total of nearly 200 million acres (> 300,000 square miles), two-thirds of which are national forests.

The idea of national wildlife refuges began early. In 1868, when President Ulysses S. Grant designated the Pribilof Islands as a northern fur seal reserve, it marked the first effort to set aside federal land specifically for wildlife conservation. These Bering Sea islands supported the world's largest fur seal rookery, and commercial interests—potential revenue from sustainable take—motivated this action. Two decades later, President Benjamin Harrison used the Forest Reserve Act of March 3, 1891,[67] to establish the Afognak Island Forest and Fish Culture Reserve.[68]

By the end of the nineteenth century, as a consequence of the aggressive post–Civil War approach to nation building and industrialization, a number of commercially viable species became threatened, including bison, wading birds, and passenger pigeons. Public opinion, especially in the big northeastern cities, demanded action to reverse these trends. In response, sportsmen's organizations and scientific societies began lobbying Congress. True to Teddy Roosevelt's legacy, the Boone and Crockett Club took the lead. This club, today known largely for assessing federal and state wildlife policies by monitoring number and size of native North American big game animals, was founded in 1887 by a group of leading explorers, writers, scientists, and political leaders, including Theodore Roosevelt and George Bird Grinnell; later Gifford Pinchot and Aldo Leopold would become members. This public outcry resonated with Roosevelt, and in response he signed an executive order on March 14, 1903, establishing the three-acre Pelican Island bird reservation along Florida's central Atlantic coast, to protect its brown pelican population. Pelican Island is considered the first bona fide "refuge." The Audubon Society paid its warden, Paul Kroegel, one dollar per month. Following this precedent, other areas—including Breton, Louisiana (1904), Passage Key, Florida (1905), Shell Keys, Louisiana (1907), and Key West, Florida (1908)—were set aside for the protection of various species of communal nesting birds. In 1905 the Biological Survey was charged with managing these reservations. By the end of his administration, Roosevelt had issued fifty-one executive orders establishing wildlife sanctuaries in seventeen states and three territories.

In addition, Congress established the Wichita Mountains Forest and Game Preserve in Oklahoma (1905) and the National Bison Range in Montana (1908)

FIGURE 44. William Temple Hornaday with the young bison he brought back to the National Zoo. Used with permission of the Smithsonian Institution. Smithsonian Institution Archives, image #74-12338.

for William Temple Hornaday's bison propagation plans.[69] When, in 1908, the Izaak Walton League helped create the National Elk Refuge in Wyoming by purchasing lands that they then donated to the government, it became the first sanctuary referred to as a refuge. In 1913 an additional 2.7 million acres were set aside when President William Taft designated the Aleutian Island chain a refuge.

In response to the growing threat to Pacific coast seabird populations due to overharvesting of eggs and feathers, three more areas—Quillayute Needles in Washington (1907), the Farallon Islands in California (1909), and areas of the Hawai'ian Islands (1909)—were given federal reserve status, which began the practice of creating wildlife refuges on Bureau of Reclamation sites.

With these scattered acquisitions, many felt it was time to formalize a national wildlife refuge program. In response to the passage of the Weeks-McLean Migratory Bird Law of 1913 and the Migratory Bird Treaty Act of 1918,[70] in 1921 two Republicans, Senator Harry New of Indiana and Representative Daniel Anthony of Kansas, introduced a migratory bird refuge bill into their respective legislative branches. There were conflicting ideas about what refuges should be. While Survey Chief Edward W. Nelson was calling for waterfowl refuges in the strict sense of the word "refuge," the American Game Protective Association was arguing for hunting opportunities. To strengthen their argument, they suggested a hunting license fee to help pay for land acquisition. Will Dilg—who founded the Izaak Walton League and pushed for the passage of the 1924 Upper Mississippi Refuge Act—fought back: "I do not care anything about saving game and fish for sportsmen. I want to save something of vanishing America. For its own sake!" Hornaday sided with Dilg and the anti-hunters, and not so subtly pressured Nelson to reduce waterfowl bag limits.

After an eight-year battle (the 1921 bill in various forms was introduced and rejected four times), in 1929 the refuge bill passed and became the Migratory Bird Conservation Act. The costs for managing and expanding the system were initially funded by congressional appropriations. With the passage in 1934 of the Beck Committee's Migratory Bird Hunting and Conservation Stamp Act (see above, known colloquially as the Duck Stamp Act), the burden of funding the refuge program shifted to revenues from the sale of duck stamps. This legislation also permitted a part of a refuge's area to be opened to waterfowl hunting (an area now set as not exceeding 40 percent of any refuge by the National Wildlife Refuge System [NWRS] Administration Act of 1966). The first acquisitions targeting waterfowl management were Dilg's Upper Mississippi River Wildlife and Fish Refuge (1924) and the Bear River Migratory Bird Refuge (1928). In 1935 Ding Darling was appointed head of the Bureau of Biological Survey and brought with him J. Clark Salyer to manage and expand the refuge system. For then until his death in 1966, Salyer was the driving force behind acquiring and administering refuges. Today, the National Wildlife Refuge System comprises more than 520 units in all fifty states, American Samoa, Puerto Rico, the Virgin Islands, the Johnson Atoll, Midway Atoll, and several other Pacific Islands. Refuges now encompass over 93 million acres of critical wildlife habitat.

Then there are the wilderness areas. The Wilderness Act of 1964 is notable not only for what it accomplished, but for the depth and quality of its phrasing, a by-product of Howard Zahniser's fine mind. As proof, I offer his definition of wilderness, as it appeared in the bill:

> A wilderness, in contrast with those areas where man and his own works dominate the landscape, is hereby recognized as an area where the earth and its community of life are untrammeled by man, where man himself is a visitor who does not remain.[71]

The designation of wilderness areas first began during the late teens and early 1920s in a handful of remote western and northern regions including Aldo Leopold's Gila Forest in New Mexico, Arthur Carhart's Trappers Lake in Colorado, and Ernest Oberholtzer's Boundary Waters in northern Minnesota.[72] "Ober" had an undergraduate degree from Harvard but was more interested in Ojibwa history and traditions than a steady job. He and an Ojibwa friend explored the abandoned Hudson's Bay outposts of the region.[73] The national effort to preserve wilderness originated, ironically, to counter sincere efforts by the federal government to conserve land.

Stephen Mather, first director of the National Park Service, wished his distant parks to serve motorized tourists, so he and his immediate successors embarked on hugely successful national campaigns encouraging Americans to "See America First" by visiting their "national [park] playgrounds." To accommodate this increase in visitation and provide more of what tourists came to see, the Park Service began constructing roads into some of the nation's most scenic and inaccessible landscapes.[74] These activities accelerated after Franklin Delano Roosevelt assumed office in 1933.[75] As a part of his work relief effort, Roosevelt created programs such as the Civilian Conservation Corps (CCC), based on the notion that work in nature addressing tangible conservation needs would not only strengthen America's infrastructure, but also heal its soul.[76]

The national parks immediately felt the impact of New Deal programs, which sent $17 million to the Park Service for road and trail development.[77] The Forest Service also received massive New Deal funding and used CCC laborers to build new roads in order to, among other things, improve firefighting access. This Park Service/Forest Service commitment to road construction alarmed preservationists. By building roads, these agencies were destroying the landscapes they were by law supposed to protect.[78]

From the beginning, the Park Service cooperated with railroads, tourist bureaus, chambers of commerce, automobile associations, and other parties interested in incorporating national parks into their programs.[79] The first automobile entered Yellowstone on August 1, 1915; five years later, there were more than 13,000.[80] Between 1915 and 1920, annual national park visitation increased from 334,799 to 919,504.[81]

Robert Sterling Yard, who would become one of the founders of the Wilderness Society, directed the Park Service's Educational Division. To support his effort, he formed the National Parks Education Committee. Soon this committee grew to over seventy members and included many of the nation's leading scientists and educators. In a May 1919 meeting held at the Cosmos Club, members formed the National Parks Association (NPA) to further their objectives, which included the promotion of national parks and education about their importance.[82]

Bob Marshall, another Wilderness Society founder, worked for the Forest Service and was also concerned with New Deal conservation policies. In 1933 he worried that the increased funding for Forest Service CCC camps would lead to "needless road building into the few wilderness areas which remain through the use of emergency funds."[83] Marshall joined the NPA and became a member-at-large of the NPA's board.[84]

On January 1, 1935, Oberholtzer, Yard, Marshall, Leopold, and four others announced the formation of the Wilderness Society.[85] Its early message was that if wilderness was to be a sanctuary in the modern world for nature untrammeled by humanity, and if it was to remain one of the last places on Earth where the primitive conditions of the frontier could be experienced at firsthand, then the intrusions of automobiles and highways must be resisted at all costs. The effects of logging and mining might be managed to keep them localized and under control, and cattle grazing, if properly regulated, might even serve as a reminder of the frontier experience, but once a road had been pushed through a wilderness area, there was almost no stopping the forces of development from eroding and finally destroying it.[86] Without these early efforts of the Wilderness Society, there would likely be no Wilderness Act or wilderness areas.[87]

Howard Zahniser was born on February 25, 1906, in Franklin, Pennsylvania. He spent his youth reading and—inspired by Nuttall, Wilson, Audubon, and Florence Merriam Bailey—developed an interest in birds.[88] Young Zahniser explored the Allegheny National Forest, established in 1923, and the Allegheny River.[89] When he was seventeen, he developed osteomyelitis and suffered two operations, conducted without anesthetic, to remove the infection.[90] As a result of this illness, Zahniser could never muster the endurance to join in the long hikes or rigorous climbs enjoyed by his family, friends, and colleagues. As an adult, his unmade bed of a face, floppy hat, and stout stature made him seem much larger than his five feet nine inches height, and deeply contrasted with the tanned, rugged appearance of colleagues such as Olaus and Adolph Murie, Ernest Oberholtzer, and Aldo Leopold.[91]

Zahniser attended Greenville College, in the Illinois town of the same name. He majored in English, minored in history, and spent four years on the debate team. After Zahniser graduated in 1928, he taught high school.[92] In 1930 he took a job as an editor for the federal Department of Commerce and moved to DC. A year later, he took a similar job with the Biological Survey, where he stayed for a decade. At the Survey, he edited manuscript drafts written by agency biologists and wrote press releases, speeches, and scripts for films. He eventually became the Survey's public relations point person.[93]

Zahniser used his Survey position to his advantage. He traveled—visiting national wildlife refuges across the country—and learned the fundamentals of ecology and wildlife management.[94] This Survey work, from the writing and editing to the interviews and long conversations with biologists and refuge managers, laid the foundation for what would become the real work of his life, managing the Wilderness Society and writing the legislation that would become the Wilderness Act (1964).

In early 1944, Zahniser met with the executive committee of the Wilderness Society at the Cosmos Club. At that time Yard, who headed the Society and was the editor of *Living Wilderness*, its chief publication, was in ill health. Committee members asked if Zahniser would be willing to edit an issue or two of *Living Wilderness* and consider a permanent position. Zahniser—with his comfortable government job—turned them down. A year later, with Yard near death, Zahniser agreed to "develop the magazine in accordance with [the committee's] ideas, consulting with and deferring to Mr. Yard."[95]

Following Yard's death on May 17, 1945, the executive committee was also looking for an executive secretary. They considered Sig Olson and Olaus Murie but felt Olson's personality would not play in DC, and Murie refused to leave his beloved Jackson Hole. Two months later the committee decided to make Murie the executive director and Zahniser the executive secretary.[96] With the move, Zahniser took a large cut in both pay and benefits, but assumed the job that would come to define him.

At the DC office of the Wilderness Society, Zahniser's days were routine: correspondence, copyediting *Living Wilderness*, and attending meetings.[97] His real work, however, was taking the writings of wilderness advocates—John Muir, Bob Marshall, and Aldo Leopold—and working their vision into the "bylaws, publications, and working principles of the Wilderness Society, and into the fabric of American culture and law."[98] In doing so, he transformed the Wilderness Society from a small group of devoted activists into a national and respected presence.[99]

In the earliest days, the Wilderness Society was supported by the Robert Marshall Wilderness Trust and fought against the development of pristine

FIGURE 45. Olaus and Adolph Murie flanking Howard Zahniser, reflecting an amazing range of talent and accomplishment. Photo used with permission of the Murie Center, Jackson Hole, Wyoming.

areas from all manner of insult.[100] The original threat was automobiles and their requirement for roads.[101] The second—the first that Zahniser faced and one of many to come following World War II—was in 1946, when a ski area was proposed to be built east of Los Angeles, in the San Gorgonio Primitive Area, which consisted of 35,365 acres within the San Bernardino National Forest. It was the first victory for the Wilderness Society, who blocked the proposal.[102]

The next challenge came from the timber industry, whose representatives argued that Olympic Peninsula forests must be harvested to meet national timber demands.[103] Another came the same year when oil and natural gas companies applied for permits to drill in the Teton Wilderness Area and the Okefenokee National Wildlife Refuge in Georgia.[104] Dams also became an issue. The Army Corps of Engineers and the Bureau of Reclamation—backed by a powerful coalition of labor unions, banks, construction firms, and government agencies—put their collective weight behind dam projects. They had inertia—gigantic, recently completed projects such as Hoover Dam on the lower Colorado River, the Bonneville and Grand Coulee Dams on the Columbia River, and Fort Peck Dam on the upper Missouri River in northeastern Montana.[105] Many Americans took pride in these massive water projects, symbolic of American Cold War strength and commitment.[106]

Then came the Echo Park Dam proposal. This dam was to be situated inside Dinosaur National Monument, on the border of Colorado and Utah.[107] Dinosaur National Monument was initially set aside by President Woodrow

Wilson in 1915 to protect a cliff containing fossil beds, but the region was also known for its spectacular scenery, and in 1938 FDR expanded its holdings to protect the Lodore and Yampa Canyons. Here, the Bureau of Reclamation decided, was an ideal dam site.[108]

There were additional considerations in favor of the dam. The Truman administration was looking for an adequate source of hydroelectric power in the region to drive the development of the hydrogen bomb.[109] Secretary of the Interior Oscar Chapman approved the dam on June 27, 1950—the same day President Truman ordered American forces into South Korea to support its accelerating conflict with North Korea.[110]

Proposals for dams in national parks and monuments worried Zahniser, who felt such disturbances violated the National Park Service Act—which required parks be preserved "unimpaired"—and weakened the entire federal park system. Zahniser viewed the 1916 Organic Act as foundational to public land policy. To build the Echo Park Dam would allow intrusion on wilderness areas and would forever undermine the Wilderness Society's work.[111]

Finally, after a five-year battle, the 1956 Colorado River Storage Project (CRSP) legislation was passed, and the Echo Park Dam project was scrapped. Toward the end of this controversy, and its echoing of Hetch Hetchy, Zahniser ensured provisions were written into the CRSP to ensure that the battle between dam development and the integrity of national parks would not continue in perpetuity. The provision relevant to Echo Park was the following: "It is the intention of Congress that no dam or reservoir constructed under the authorization of this Act shall be within any national park or monument."[112] With that, the Echo Park Dam was dead.

There were other dangers to wilderness preservation. Perhaps the most threatening was mining. The Mining Law of 1872 allowed exploration for minerals and patenting claims on public lands, whereupon the patent took priority over all other rights.[113] This law needed to be modified if wilderness areas were to be designated.

Fire protection posed another challenge. Pinchot had pronounced the destruction caused by a fire in the Priest Lake Forest Reserve near Blackfoot, Montana, "sickening."[114] However, fire is both a natural and an aboriginal process, and many of today's plant and animal species evolved in the presence of fire. Some questions arose. Should a fire sparked by lightning within a wilderness be allowed to burn? If so, how could it be kept from spreading beyond the wilderness boundary onto private land? If not, should roads be constructed to enable access for firefighting vehicles? And how could agencies ensure fire roads would subsequently never be used and therefore maintain the wilderness?[115]

As these many threats to pristine areas became defined and more numerous, it became clear to Zahniser that neither he nor the Wilderness Society had the time or money to fight them on a case-by-case basis—a national wilderness policy was needed. Indeed, when the Wilderness Society's Executive Council met in the Boundary Waters in 1947, Benton MacKaye proposed a continental system of wilderness embracing "representative units of eleven distinctive ecological zones including the Lake State Pines, Short-grass Prairie, Rocky Mountain and Northwest Conifer Forests, Coastal Swamp, Southern Appalachian, Southwest Desert, and Tundra." MacKaye's idea was a response to Leopold's observation that key ecosystems such as "large stretches of primitive prairie" had been erased from the American landscape.[116]

As Zahniser mulled these problems, he cultivated his network. During the summer of 1947, Zahniser had long conversations with Murie, whose detailed knowledge of wildlife and the ability to draw it astounded Zahniser. Their professional relationship as co-executives of the Wilderness Society initially had been based on correspondence, so being together for much of that summer deepened their friendship and solidified their partnership. Murie, for his part, came to appreciate Zahniser's capability to work long hours and his talents for administrative detail and effective communication.[117]

Back in DC, Zahniser used the Cosmos Club to develop relationships with like-minded groups and personalities, and to meet with his network of conservationists; men such as Ira Gabrielson, his onetime boss at the Fish and Wildlife Service and later director of the Wildlife Management Institute.[118] In 1947 Zahniser, Gabrielson, and Izaak Walton League leaders met at the Cosmos Club to discuss proposed legislation permitting hunting on national wildlife refuges, which at that time were sanctuaries.[119] Zahniser relished the Cosmos Club for its origins, history, and atmosphere, and regularly ate lunch there. Because the club had guest rooms, he often found himself dining there while hosting out-of-town colleagues such as David Brower, Benton MacKaye, and Sigurd Olson, who became president of the National Parks Association in the early 1950s.[120]

Zahniser's relationship with Brower was especially important. The two had met in Knoxville at a Natural Resources Council meeting in 1950 and liked each other immediately. They began corresponding in 1953, after Brower was elected executive director of the Sierra Club. Their approaches were complementary. Brower's witty, energetic, aggressive, Tenth Mountain Division demeanor contrasted with Zahniser's methodical approach. Their rapport provided a critical connection between the Wilderness Society and the Sierra Club that proved to be vital in the big wilderness battles they would come to fight.[121]

Finally, Zahniser stated the problem and the only solution he could conjure:

Let us try to be done with a wilderness preservation program made up of a sequence of overlapping emergencies, threats, and defense campaigns.... [Instead,] [l]et's make a concerted effort for a positive program that will establish an enduring system of areas where we can be at peace and not forever feel that the wilderness is a battleground![122]

The Chesapeake and Ohio Canal ran alongside the Potomac River and had once transported commerce such as coal, corn, wheat, and flour, linking Washington, DC, with markets on the other side of the Appalachian Mountains, in the greater Ohio River Valley watershed. Begun in 1828, the canal consisted of an adjacent towpath, locks for getting up and down the mountains, as well as bridges. It was two hundred miles long, took two decades to build, and cost $27 million. It was used for over a century until railroads proved more practical, and it was closed in 1924. In 1938 the federal government acquired the canal and a 230-foot right-of-way that included the canal and towpath.[123]

In 1953 a proposal to transform the canal into a parkway as an extension of the George Washington Memorial Parkway development plan was floated, and there was pushback, first by the state of Maryland, then by environmentalists.[124] On January 3, 1954, a *Washington Post* editorial supported the parkway, declaring it would make the canal available to many more hikers, campers, and fishermen. Two weeks later, a guest editorial was published in the *Post*, authored by Supreme Court Justice William O. Douglas:

The discussion concerning the construction of a parkway along the Chesapeake and Ohio Canal arouses many people. Fishermen, hunters, hikers, campers, ornithologists, and others who like to get acquainted with nature first-hand and on their own are opposed to making a highway out of this sanctuary. The stretch of 185 miles of country from Washington, D.C., to Cumberland, Md., is one of the most fascinating and picturesque in the Nation. The river and its islands are part of the charm. The cliffs, the streams, the draws, the benches and beaches, the swamps are another part. The birds and game, the blaze of color in the spring and fall, the cattails in the swamp, the blush of buds in late winter—these are also some of the glory of the place. I wish the man who wrote your editorial of January 3, 1954, approving the parkway would take time off and come with me. We would go with packs on our backs and walk the 185 miles to Cumberland. I feel that if your editor did, he would return a new man and use the power of your great editorial page to help keep this sanctuary untouched.[125]

Merlo Pusey had written the editorial, and both he and editorial page editor Robert H. Estabrook responded on January 21 with a second editorial entitled "We Accept."

News of the hike ignited conservation leaders. Where Douglas and the *Post* had expected a small group of four or so, fifty-eight people started the hike, including Olaus Murie, Harvey Broome, Bernard Frank, and Howard Zahniser—respectively, president, vice president, executive committee chairman, and executive secretary of the Wilderness Society. Sigurd F. Olson, by then president of the National Parks Association, also walked. The Wilderness Society and the Potomac Appalachian Trail Club provided logistical support for the hike. The hikers began on March 20, averaged twenty-three miles a day, and ended on March 27. Only "the nine immortals" finished the entire 185-mile route; among them were Douglas, Murie, and Broome.

The C&O Canal hike helped to convince mid-1950s America that wilderness preservation was worthwhile. Up until then, most Americans were unaware of the nation's threatened wilderness, and many viewed conservationists derogatorily as "flower-pickers" and "bird watchers." Wilderness preservation was no longer just important to a few; with the C&O Canal hike, it grew to become a national issue.[126]

Zahniser's Wilderness Bill campaign, which began two years after the C&O hike, would take eight and a half years to pass and would span three presidents.[127] Neither he nor Olaus Murie lived to see it become law. The bill proposed a system to preserve wilderness at the national level, "retaining their primeval environment of influence and remaining free from mechanized transportation." As proposed by Zahniser, this wilderness would include lands designated as wild or wilderness, as well as national forest primitive areas, and a method for adding new areas from national forests, national parks and monuments, wildlife refuges, and Indian reservations.[128] It was controversial.

What was the place of wilderness in America? Was wilderness the thing that pioneers had struggled against and finally conquered, and thus should be eliminated, or was wilderness a defining force in the nation's history that must be preserved as a reminder of the pioneers' struggles? Further, couldn't wilderness serve as a unique source of adventure, giving, as many have argued, a spiritual centering to an increasingly overtaxed and stressed urban society?[129]

While campaigning for the presidency, John F. Kennedy endorsed the Wilderness Bill as part of his "New Frontier." Soon after his inauguration, in February 1961, the new president gave a special address on natural resources to Congress and urged the Wilderness Bill's passage. Kennedy's stance obliged

Stewart Udall, the new secretary of the Interior, and Orville Freeman, the new chief of the Forest Service, to align their agencies accordingly.[130] It took another three years. The Senate passed the Wilderness Bill for the second time in April 1963.[131] Olaus Murie died on October 21, 1963; Howard Zahniser died on May 5, 1964. Two months later, the House finally passed the Wilderness Bill by a vote of 374–1. On September 3, 1964, President Lyndon Johnson signed the bill into law during a special ceremony held in the Rose Garden.[132] The act states that wilderness areas "shall be administered for the use and enjoyment of the American people in such a manner as will leave them unimpaired for future use and enjoyment."[133]

In recent years, a subset of environmental activists have pointed out the shortcomings of the National Wilderness Preservation System. To their way of thinking, wilderness areas were not sufficient by themselves, for they too often stood as isolated patches within a larger matrix of agricultural, logged, and urbanized landscapes. Most wilderness areas provided too little room for large, wide-ranging animals such as grizzly bears, cougars, and wolves. Only by connecting wilderness preserves, national parks, national forests, wildlife refuges, and other protected lands with habitat corridors could mobile species be expected to survive and biological diversity enhanced. Despite this view, even influential figures such as Dave Foreman, architect of the Wildlands Project, appreciated the National Wilderness Preservation System as a crucial building block leading to current efforts.[134]

Perhaps the most eloquent critique of the idea of wilderness designation came from Wendell Berry, who is always eloquent:

> You can't save wilderness preserves, refuges, and national parks, if at the same time you let the economic landscapes and the land-using economies go to the devil. I can't look at the crisis of the Arctic Wildlife Refuge except as the result of a radical failure of the conservation movement over the last fifty or so years: its refusal to see that conservation as we have known it is not an adequate response to an economy that is inherently wasteful and destructive; its apparent belief that nature or wilderness can be preserved merely by preserving wilderness; its inability to connect wilderness conservation with soil conservation or energy conservation or any form of frugality; its cherished contempt for ranchers, farmers, loggers, and other land users. . . . [C]onservationists . . . have cared too little for landscapes that were not describable as "wilderness" or "open space." They have too thoughtlessly "benefitted" from cheap food and cheap fuel. When I think of the threat to the Arctic Wildlife Refuge, I think also of conservationists and wilderness lovers who drive thousands of miles to walk a few hours or days in a certified wilderness. We have got to think of something better. If we don't, the government won't.[135]

The Sole Midwestern Wilderness: Quetico-Superior
Boundary Waters

In addition to the Midwest producing naturalists, ecologists, wildlife biologists, and conservation biologists who deeply influenced the United States' policies and direction on national parks, national forests, wildlife refuges, and wilderness areas, the Midwest offered a region, the Boundary Waters, that in a unique binational cooperation would come to define the essence of wilderness. Indeed, the first two pieces of federal legislation anchoring wilderness protection were passed in response to Boundary Waters challenges. The 1930 Shipstead-Nolan Act banned shoreline logging in areas of the Boundary Waters, while the 1948 Thye-Blatnik Bill authorized the Forest Service to buy land for recreation.

In 1926 the Forest Service designated a portion of the vast Superior National Forest as wilderness. Eight years later, FDR formed the Quetico-Superior Committee to explore the idea of creating an international preserve linking Minnesota's canoe country with the vast lake system in Canada's Quetico Provincial Park, an idea similar to the preserve formed in 1932 linking the American Glacier and Canadian Waterton National Parks (Waterton-Glacier International Peace Park), situated along the Rocky Mountain/Great Plains ecotone divide. Roosevelt asked local activists, including Ernest Oberholtzer, to serve and requested the Departments of Interior and Agriculture to name representatives; Interior's Harold Ickes chose Bob Marshall. Leopold, Yard, and Mac-Kaye served as advisors. They met at Oberholtzer's cabin on Mallard Island in Rainy Lake and began sharing ideas about a modern wilderness policy.[136]

Twenty years later, in June 1947, the Wilderness Society met at Oberholtzer's cabin, and MacKaye proposed a continental system of wilderness embracing "representative units of eleven distinctive ecological zones including the Lake State Pines, Short-grass Prairie, Rocky Mountain and Northwest Conifer Forests, Coastal Swamp, Southern Appalachian, Southwest Desert, and Tundra." The idea was based on Leopold's point that key ecological niches such as "large stretches of primitive prairie" had been erased from the American landscape. Preserving the Boundary Waters was deemed "central to this program."[137] "This program" would become the Wilderness Act of 1964.

Despite Oberholtzer's priority and early influence, today no one is more associated with the preservation of the Boundary Waters than Sigurd Olson. Born in Chicago on April 4, 1899, seven-year-old Olson moved north with his family to Sister Bay, in Door County, Wisconsin. It was there, as a child, where he first heard the "singing wilderness."[138] Reminiscent of John Muir's father, Olson's father, a Swedish Baptist minister, was formal, rigid, and grim.

He once discovered a chess set Olson and his brother had bought, got angry, and threw it into the fire.[139] When Olson was ten, his father became the "state missionary for the Swedish Baptist State Convention." The family moved to northwestern Wisconsin—first to Prentice, then to Ashland.[140]

By then, the old-growth white pines that historically dominated the region had been clear-cut, and the landscape converted from Northwoods forest to the agriculture frontier of "stump farmers."[141] Sig Olson needed to be outdoors; these were his "Daniel Boone days."[142] As with Teddy Roosevelt and other biologists profiled here, Olson read adventure novels, hunted, and trapped.[143] Olson's interests were not his family's interests, and they separated him from them, although his grandmother, "Mormor," understood.[144]

Olson entered high school at age thirteen, which even then was young, and won a five-dollar gold piece for writing an essay on "The Function of the Chamber of Commerce."[145] He graduated from high school in 1916 and entered Northland College, a two-year, evangelically rooted prep school.[146] At Northland, Olson befriended Andrew Uhrenholdt, the fourth child of Søren Uhrenholdt. The elder Uhrenholdt was a gifted man who had emigrated from Denmark.[147] He had bought 160 acres along the Namekagon River at Seeley and began stump farming.[148] In 1903 the Wisconsin legislature established a state forest system with the goal of reforesting the cutover and creating a sustainable logging industry. However, the northerners wanted agriculture and fought the idea, eventually winning—in 1915 Wisconsin courts declared the forest system illegal.[149] It didn't matter to Søren Uhrenholdt, who had never bought into the perception of agriculture and forestry as mutually exclusive endeavors.[150]

Olson worked summers at the Uhrenholdt farm. He learned to fly-fish, shoot, sing songs, and tell stories,[151] and he began to court his future wife, Elizabeth Uhrenholdt—Søren's daughter and Andrew's younger sister. In the fall of 1918, Olson enrolled at the University of Wisconsin and joined a missionary organization, although his missionary work may have had more to do with being outdoors than saving souls.[152] Indeed, in Madison, Olson began losing his religion[153] but remained spiritual. In 1920, after graduating, he was offered a job teaching high school in the heart of the Mesabi Iron Ore Range, north and west of Duluth. After the Northwoods were clear-cut, the iron ore and copper mining industries drove the economies of Michigan and Minnesota, respectively,[154] and were of national importance. (They are relevant to our story, here, too. Following the Civil War, Alexander Agassiz, son of Louis, invested in and then managed the Hecla Copper Mine, in Calumet, Michigan, in the Upper Peninsula. As a result of his efforts and his investments, Agassiz and his family became wealthy, and their newfound prosperity helped to complete Harvard's Museum of Comparative Zoology.[155])

On the Iron Range, Olson taught school weekdays and hiked and camped on weekends. About school teaching, he said, "I had no professional pride in what I was supposed to do, no sense of mission. I would do what was required of me, but that was all. . . . In town I had a job to do, but [in the woods] was where I really lived."[156] Olson became obsessed with wilderness.[157] While hiking through the woods, he came across wolves eating a deer carcass, and it made a deep and disturbing impression on him.[158]

Olson met outdoorsman Al Kennedy, who had been pardoned by Teddy Roosevelt after being jailed for murder. Kennedy invited Olson to go with him to the "Flin Flon" region in Manitoba for two or three years to pan for gold. Elizabeth, at this point already exasperated with Sig's wilderness wanderings, gave him an ultimatum—that if he went, she would not wait for his return.[159] Olson stayed, but he followed Kennedy's suggestion to explore the canoe country along the Minnesota/Ontario border. When Olson arrived, in the summer of 1921, portions of the American side were clear-cut.[160] When he encountered old growth, he thought, "This is beautiful. Everything as God made it, untouched by man."[161]

The Olsons were married on August 8, 1921, but Olson was a wayward husband (not in the usual sense—I mean he spent more time in the outdoors than with his wife).[162] As his preferences came into sharper focus, he realized writing was more fulfilling than teaching and began to contemplate writing fiction.[163] It would be a while before he began, much longer before he became successful.

In the 1920s, the University of Wisconsin allowed anyone with their bachelor's degree to enter graduate school. In the fall of 1921, Olson enrolled in the Geology Department. In those days Wisconsin hosted a top-five program in geology, ranking with Harvard, Yale, Columbia, and the University of Chicago.[164] In the Geology Department, Olson gained an appreciation for the Boundary Waters' 2.5-billion-year-old Precambrian rock and learned to appreciate Earth as a living, dynamic entity. Then, just as quickly as he had entered graduate school, he left, disenchanted with academics and his separation from Elizabeth, who was pregnant, and his beloved Northwoods.[165] His in-laws became increasingly worried: "We don't think so much 'bout that quitting and moving around."[166]

Before Olson left Madison, he impressed the superintendent of Ely's public schools and landed his second teaching position. The Olsons arrived in Ely in 1923, and students found Sig a fair and friendly instructor.[167] He took field trips because he thought biology and geology could not be taught adequately without firsthand experience.[168]

Olson's wilderness interests led to summer work outfitting and guiding. Although Olson never fit in, he greatly admired the older guides and learned many outdoor skills from them. Several were descendants of the voyageurs and shared old stories.[169] When not guiding, these men worked as lumberjacks, trappers, or miners. In town, they suffered low social status but "didn't much care."[170]

Sig loved *The Voyageur* by William Henry Drummond;[171] indeed, he loved the whole idea of the voyageurs. He would, on occasion, dress in their traditional costume of red blouse, breechcloth, moccasins, sash, and toque. Olson would have been tall for a voyageur, who rarely topped five foot five.[172]

Olson developed a loyal clientele,[173] and his lifestyle of alternating summer guiding with his school-year obligations persisted through the 1920s. His academic responsibilities shifted in 1926, when he began teaching at Ely Junior (now Vermillion Community) College.[174] He began writing stories he sent to magazines such as *Harper's* and the *Saturday Evening Post*. They didn't sell, although in 1927 *Field and Stream* published "Fishing Jewelry," and in the next two years he got three more articles published.[175] During the late 1920s, Olson became a partner in the Border Lakes Outfitting Company and reduced his guiding.

Olson was obsessed with finding the meaning of his life—a calling that reflected a higher purpose.[176] At one point, he climbed a Northwoods peak at sunset and "felt a sense of peace and contentment, and a belief that he had experienced contact with a supernatural power." From that moment on, he claimed to be a spiritualist, and he began seeking these epiphanies, or "flashes of insight": he said he would "find what I sought on brilliant starlit nights."[177]

In 1924 Olson guided Izaak Walton League founder, Will Dilg. During this trip, Dilg promised that the Ikes would fight to preserve the canoe region. Olson joined the League and in 1926 participated in the first big battle over the preservation of these boundary waters. Superior National Forest had been established in 1909. After Aldo Leopold had convinced the Forest Service and the Secretary of Agriculture to set aside 574,000 acres of the Gila National Forest as a wilderness area,[178] the Ikes wanted similar protection for Superior. In particular, they wanted the federal government to purchase private lands interrupting national forest holdings to create a large contiguous wilderness area.

Business leaders were angered by the prospect of federal control of their lakeshores. In their eyes, tourism was the only chance to develop and nurture a locally controlled commerce. Their historic economy had been logging, and once the region had been clear-cut, the timber industry abandoned it. The mining industry was not much better. It had busted unions by importing cheap eastern European labor, eliminated jobs through mechanical advances, and declared layoffs on short notice. Small businesses centered on tourism

FIGURE 46. Sig Olson demonstrating the practical clothing of a Boundary Waters guide during a 1955 trip to the Upper Churchill River. Image used with permission of the Wisconsin Historical Society, image #WHS74064.

offered an alternative. The *Ely Miner* was a pro-business newspaper and called the Ikes' demands for extensive wilderness "foreign to the principles of Democracy and striking at the very root of Americanism."[179]

Local Forest Service managers were also opposed. They wanted roads built to facilitate moving people and equipment on forest fires. But as always, roads brought in developers. The outfitters and guides knew roads would bisect major canoe portaging areas and open immense sections of land to resort development that would destroy any sense of seclusion on affected lakes.[180] In September 1926, Aldo Leopold brokered a compromise. At that time, Leopold was assistant director of the Service's Forest Products Laboratory in Madison. He knew the Quetico-Superior region from two canoe trips he had taken. Leopold persuaded the Ikes' national office that it was impractical to preserve the entire American portion as wilderness. Instead, better to compromise and preserve some of it now, rather than risk losing it all through inflexibility.

In 1928 Olson guided University of Illinois ecologist Alvin Cahn. Olson later said, "It was through him that I first caught the meaning of ecology as a concept, and now as I look back one thing stands out, the discovery of the sense of interdependency of all living things." After the trip, Cahn began mentoring Olson, broadening his education by sending him books on history, world affairs, and literature, and urging him to consider graduate school.

Olson was not enthusiastic. He understood himself well enough to know he preferred writing fiction to doing science. As he read Thoreau, John Burroughs, and William Henry Hudson, he concluded that these authors were saying and doing things closer to what we wished to say and do—"akin to me in every action, every thought"—than were the ecologists.[181] He followed up on these thoughts with action, sending a story to a literary agent, who criticized its character development but noted that Olson wrote "effectively about [the] life and activity of [the] region."

In 1930 Olson reversed himself and wrote to Cahn about graduate school. Cahn enthusiastically replied, "I had a nice talk with [Victor] Shelford this week about you, and he is thoroughly delighted with the prospect of having a graduate student interested in vertebrate ecology—a very rare combination." In preparation for his master's thesis, Olson gathered information on the big predators of Superior National Forest.[182] As he began his research, in December 1930 he published "The Poison Trail" in *Sports Afield*. The article detailed a winter trip through the Boundary Waters to pick up carnivores poisoned by Biological Survey trappers, whose main target was timber wolves. They traveled about a hundred miles through "one of the finest big game areas on the continent and incidentally one of the most harassed by the killing packs," and collected, in total, the carcasses of ten fox, four coyotes, and four wolves.[183]

At this stage, Olson showed little understanding of predators. He called wolves "grey marauders" and "killers," and their predations were "murders." When he found a wolf-killed deer carcass, he exclaimed, "What we most feared, had happened. Only the fat of the entrails had been eaten. [The wolves] were already killing for fun." These feelings were not just reserved for wolves but extended to all predators. He called the elegant snowy owl the "most feared of all the winged marauders of the North" and described "the swish of his great feathered pinions . . . an omen synonymous with tragedy."[184]

At Illinois, Olson got on with Shelford but despised indoor laboratories, the Latinized species names he had to memorize, and the jargon of science. "I hate the very sound of the word Ecology. . . . When I get through with it here, I am through. I will bury it away someplace and never again as long as I live ever touch it."[185]

During his final semester at Illinois, Olson worked on his writing, completing class-required papers on ecological succession in ecological communities such as prairies, hardwood forests, lakes, and—Shelford's area of specialty—Indiana Dunes. Olson published two sections of his thesis on predation in the journals *Ecology* and *Scientific Monthly*.[186] About Olson's work, wolf expert Dave Mech commented, "Although it did not lead to any major insight into the life of the wolf, it was a milestone in first treating the wolf as a worthy subject for scientific research."[187] Olson's experience in Champaign also represented an epiphany in his thinking about wolves. Under the influence of Shelford and Cahn, Olson finally understood the role of predators in natural, healthy ecosystems. With his master's degree in hand, by the summer of 1932, Olson was back in Ely.[188]

In early November 1933, Olson received a letter from Aldo Leopold, who in July had accepted an appointment from the University of Wisconsin as the world's first professor of game management. He was looking for graduate students: "I have no funds as yet, but if I had an extra strong man to line up and a definite amount to shoot at, it might help me to get them. . . . What would be the minimum stipend necessary to attract you?" Sig was stuck. To be pursued by Leopold was big.[189] On the flip side, to turn Leopold down "would forever damn me in his eyes and in the eyes of all who are interested in game."[190] After agonizing about this decision, he finally wrote:

> I am afraid that I will have to ask too much [for a stipend] to put me in the running. I am 34 years of age, have a family of two growing boys to support, insurance, and other incidentals which would make it impossible for me, in

view of the fact that my annual outlay has been keyed up to a certain figure, to accept a stipend as low as I could were I single and on my own.[191]

In the fall of 1933, Olson traveled to Minneapolis to testify at a public hearing on the management of the Quetico-Superior canoe country. Entrepreneur Edward Backus had planned a massive dam-building campaign in the United States and Canada between Lake Winnipeg and Rainy Lake. The purpose was to generate electricity; costs were to be covered by customers and the governments of the United States and Canada. Water levels in boundary waters lakes would rise by as much as eighty feet, thousands of miles of shoreline would be flooded, and extensive shoreline stands of virgin white pines would be inundated.[192] After his testimony, Olson thought, "Somehow I have the power of conveying my enthusiasm to others, particularly men. I can make them see and feel what I see and feel of the out-of-doors."[193] Not only that, Olson's testimony got the attention of Ernest Oberholtzer, who had connections among the nation's most prominent conservationists.[194]

American conservationists asked Congress for legislation forbidding any alteration of water levels on federal land and a ban on shoreline logging. They got it. When former Ike officer President Herbert Hoover signed the Shipstead-Nolan Act into law on July 10, 1930, it was the first time the U.S. government enacted a law designed to preserve wilderness.

In the fall of 1934, Olson taught only at Ely Junior College,[195] and with his less-complicated schedule, he began once again to consider becoming a professional writer. He had his hook. "No one has yet developed a philosophy of the wilderness. That is up to me. . . . My work must be strong and hard and masculine, the love of men for the world, its truth, its unvarnished joy, its compensations, the feeling of being alone."[196] On New Year's Day, 1935, Olson moved out to a cabin ten miles south of Ely, took a copy of Thoreau's *Walden*, tried to write, but couldn't. Then, about six weeks later, he got an idea and wrote his first wilderness essay, "Farewell to Saganaga," in a first-person, narrative style, which would become his trademark.[197]

In 1935, the year of the Dust Bowl, Olson took advantage of several opportunities. In March, Shelford asked him to chair a committee of the Ecological Society of America on reintroducing wolves to Isle Royale. Then in May, Bob Marshall, director of the U.S. Office of Indian Affairs, asked Olson to visit Indian reservations in northern Minnesota and Wisconsin to explore ways to increase tribal income through recreational development.[198] Olson had guided Marshall the summer before, and on January 1, the same time Olson has heading out to the cabin to write, Marshall—along with Oberholtzer, Leopold, Robert Sterling Yard, Harvey Broome, Harold Anderson, and Benton MacKaye—published a pamphlet announcing an organization to be known as the Wilderness Society, "for the purpose of fighting off invasion of the wilderness and of stimulating . . . an appreciation of its multiform emotional, intellectual, and scientific values."[199]

A year later, in January, Olson became the dean of Ely Junior College. It didn't take long for Olson to realize that his new position required him to be "on call" 24/7, with no free time to write. There was good news. That year, Franklin Delano Roosevelt added 1.3 million acres to Quetico-Superior, which made Superior the nation's largest national forest.[200]

In 1937 Marshall brokered a deal to stop the Fernberg Road to Basswood Lake.[201] Later that summer, Olson and Marshall spent five days canoeing Quetico-Superior.[202] During this period, Olson developed his philosophy of wilderness—the key to his argument for preserving it: "Because man's subconscious is steeped in the primitive, looking to the wilderness actually means a coming home to him."[203]

In 1940 Olson did something that a decade later he would question and perhaps regret—he introduced smallmouth bass to the Quetico-Superior.[204] From our modern perspective, we might assume Olson's doubt was due to the negative ecosystem effects of introducing a new top predator to a pristine wilderness lake system (i.e., a lake with intentionally introduced species does not represent "untrammeled wilderness"). But it was not. Instead, Olson noted the presence of bass made the Boundary Waters more attractive to fishermen, who flocked to the region, increasing the human footprint.[205]

During the fall of 1942, Olson's son Sig Jr. enlisted in the army and was chosen for service in the famed Tenth Mountain Division. In 1944 his unit experienced sustained heavy fighting in northern Italy. Despite suffering 75 percent casualties, Sig Jr. was unharmed. Martin Murie, the first son of Olaus and Mardy Murie, was also in the Tenth Mountain and also survived.[206]

Immediately following the Second World War, Sig Sr. went to England to teach idle GIs waiting deactivation. There, he dined with famed ecologist Charles Elton, who had read his wolf research.[207] He stayed in Europe through June 1946, where he toured twenty-two concentration camps, including Dachau and Bergen-Belsen. He attended the Nuremberg trials as an observer for the State Department. Exposure to Nazi atrocities and the men who perpetrated them shook his foundation enough to be receptive to the encouragement he received from professional ecologists and to the work of Pyotr Alekseyevich Kropotkin. From Kropotkin, Olson learned that he should abandon writing adventure stories and instead write about communicating with "the cosmos and man"—"[to] go into the wild and come out with a message."[208]

After the war, and for much of the rest of his long life, Olson was plagued with nagging stress-related health issues. He suffered from ulcers, his eyes were painfully sensitive to light, he became an insomniac, and in 1947, during his final year as dean at Ely Junior College, he developed a facial tic.[209]

After resigning his deanship, Olson accepted a job as the Ikes' wilderness ecologist and advisor to the President's Committee and the Wilderness Society.[210] He acquired a literary agent and began writing. Both careers had rocky beginnings. About Olson's position with the Ikes, Charles Kelly of the Quetico-Superior Council volunteered, "I have never wanted Sig for our program because he is a little too fanatic."[211] And, after reading one manuscript, a reviewer offered, "The writing is not adequate. The cliché and the trite (word and phrase) are too often found. Errors of techniques harm the effect; the transitions are abrupt. He makes punctuation and other [esp. spelling] mistakes that irritate editors."[212] Olson was nothing if not persistent.

In person, and while public speaking, Olson had the grace, poise, and confidence that being comfortable in wilderness brings to people. He was charismatic and, in the manner of an accomplished singer, had a strong, attractive voice.[213] He knew what he was talking about in the field. He could discuss the age, composition, and formation of rocks, their lichens, the composition of forests, their bird and mammal species, and patterns of ecological succession. He could also talk about the human history of the Northwoods and the voyageurs. As one journalist noted, Olson "could look at a lake and know right where to go to get a couple of fast lake trout for supper while the rest of us pitched tents. . . . In that two weeks [when he guided me,] Sig Olson assumed for me a kind of mythic godlike caliber that mellowed a little as my experience broadened but never entirely left me."[214]

Olson's star rose in the early 1950s. He was invited to speak at Sierra Club and American Wildlife Institute meetings, was inducted into the Cosmos Club, and became the president of the National Parks Association (NPA).[215] Thanks to his charisma, the NPA would come to have more members than the Wilderness Society.[216] In 1968 Olson became the president of the Wilderness Society, a position he held for three years.[217]

Olson played a role in many of the major conservation fights from the 1950s to the mid-1970s, although he was never considered a political infighter.[218] Olson was also writing. His break came when conservation groups met on November 17, 1954, at the Baroque Suite of the Plaza Hotel, to discuss strategy after Eisenhower approved MacKaye's proposal to move forward on the Echo Park Dam project.[219] Olson gave the keynote. The publisher Alfred A. Knopf was in the audience and came away deeply impressed.[220] Knopf wrote to Olson, "I am wondering if you are not going to have a book for us one of these days?"[221] Olson had a manuscript nearly ready, and in 1956 Knopf published his first book, *The Singing Wilderness*. In reviewing the book for the *Milwaukee Journal*, Gordon MacQuarrie pronounced it an "inspiring book . . . of Leopold quality." Roger Tory Peterson wrote that it was "unequivocally the best series of essays on the northwoods country I have ever read."[222] The book was a commercial success and made the *New York Times'* Best Seller list.[223] In 1956 and 1957, Olson began working on a series of new essays, which became his second book, *Listening Point*, published in 1958, which David Backes describes as Olson's search for a land ethic.[224] In total, Olson would publish nine books, including his autobiography, *Open Horizons*.[225]

At his core, Olson was a wilderness theologian[226] who felt an intellectual, generally unacknowledged, descent from the Transcendentalists.[227] He believed that "the spirit of man is the flowering of nature, greater than any other phenomenon, greater than the whirling spheres, greater than space, infinity." Consistent with this notion, Olson's mistrust of science was that it is based only on rational analysis, and therefore offers an incomplete view of nature. He felt his mission in life was "to give men a picture of God [nature] as he really is."[228]

In 1974 Olson received the John Burroughs Medal, America's highest award for nature writing. Eight years later, while snowshoeing, he had a fatal heart attack. In his typewriter was a sheet of paper with these words in the following arrangement:

> A New Adventure is coming up
> And I'm sure it will be
> A good one.

Two Agencies Face Tough Transitions

The utilitarian conservation/preservation battle personified by Pinchot and Muir was about more than land designation within federal agencies—it was also about the policies being carried out by these agencies. During the twentieth century, both the Bureau of Biological Survey and the Forest Service shifted from a conservationist philosophy of hands-on natural resource control—"the greatest good for the greatest number"—to a hands-off preservationist philosophy of "let nature take its course," what Nash has called the greening of philosophy.[229] These were painful changes. The lessons learned from the history of these agencies over the course of the twentieth century taken together symbolize the shifting attitudes of Americans toward the treatment of the natural history of their country.

BUREAU OF BIOLOGICAL SURVEY/U.S. FISH AND WILDLIFE SERVICE

As early as the 1880s, western ranchers appealed for state and federal assistance to control predators such as gray (buffalo) wolves, mountain lions, grizzly bears, coyotes (plains wolves), bobcats, lynx, wolverines, fox, hawks, eagles, as well as nuisance rodents. Ranchers longed for an idealized landscape made permanently safe for "good" animals and commercial enterprise.[230]

Early predator control relied on strychnine and bounties. A single strychnine-laced buffalo carcass could kill "a hundred or more wolves," according to Montana rancher Granville Stuart.[231] Other accounts confirmed the astonishing toll this poison took, not only on wolves, but also badgers, bears, bobcats, coyotes, eagles, fox, hawks, magpies, ravens, and skunks.[232] In the mid-1850s, the owners of a trading post/ranch along the Santa Fe Trail in southwestern Kansas benefited from the exploitation of both bison and wolves. One of the operators recalled the technique. To bait wolves, they killed a bison, diced the meat, and scattered it. A couple days later, they replaced the bait with strychnine-laced meat. The operator claimed they killed sixty-four wolves in one night and earned $4,000 in profits from one winter's work.[233]

Bounties supplemented poisoning. In the 1880s, the federal government paid twenty-five dollars a head for mountain lions.[234] North Dakota paid three dollars each on wolf scalps, with ranchers kicking in another five dollars. Montana offered eight dollars per animal for bears and mountain lions. During the first year of Montana's bounty, 5,450 wolves, 565 bears, 146 lions, and 1,774 coyotes were killed.[235]

By 1900, 30 million sheep grazed throughout the western range,[236] and there were louder calls to eradicate predators. The federal government, being a major landowner in the West, was pressured to comply. In response, in 1905 the Forest Service engaged Survey biologist and C. Hart Merriam's brother-in-law Vernon Bailey to first discover where wolves lived, then hire trappers to kill them. Bailey was available because Merriam saw a chance to respond to congressmen who claimed the Biological Survey was doing nothing practical. Merriam stated that predators "levy a heavy tax on stock . . . a loss aggregating several millions of dollars annually."[237]

By the time federal predator control work started in 1905, poisoning and bounty hunting had reduced the larger predators—wolves, grizzly bears, and

FIGURE 47. Custer Wolf, killed by Bureau of Biological Survey agents, along with a sizable take of Great Plains coyotes. Photo reproduced with permission of the Arizona Historical Society, image PC10F417_5929.

mountain lions—to remnant populations. In 1907 alone, nearly 2,000 wolves and 23,000 coyotes were killed in national forests and on other federal lands in the West. Federal authorities also arranged cooperative programs with ranchers and claimed success in killing wolf pups in dens.[238] Holdout "renegade" wolves, including "Old Whitey," "The Traveler," "Old Three Toes," the "Custer Wolf," and the "Split Rock Wolf" became legends.[239]

After the First World War, the Survey's Predator Animal and Rodent Control (PARC) expanded. Between 1915 and 1924, agents killed more than 1,200 mountain lions.[240] When "Old Three Toes" was killed in South Dakota in 1923, wolves were officially extirpated from the Great Plains. Coyotes, previously ignored, became the new threat to western livestock.[241] Sheep ranchers' wish for a coyote-free range met the Survey's need for clients, so PARC agents, wholeheartedly supported by the National Woolgrowers' Association, set out to kill as many as possible. In 1924 the Biological Survey estimated that 134,000 coyotes were killed. The body count ranged between 35,000 and 40,000 per year throughout the 1920s.[242]

The Survey's program of predator control was endorsed by young Aldo Leopold, working for the Forest Service in New Mexico. In 1915, in the inaugural issue of *The Pine Cone*, Leopold encouraged cooperative efforts between hunters, game protection associations, livestock groups, and the Survey.

It is well known that predatory animals are continuing to eat the cream off the stock grower's profits, and it hardly needs to be argued that, with our game supply as low as it is, a reduction in the predatory animal population is bound to help the situation. . . . Whatever may have been the value of the work accomplished by bounty systems, poisoning, and trapping, individual or governmental, that fact remains that varmints continue to thrive and the reduction can be accomplished only by means of a practical, vigorous, and comprehensive plan of action.[243]

If stock growers and hunters could put aside their differences and combine to contribute toward more effective control, everyone, "except the varmints," would benefit mightily.

Two years later, in a 1917 address, Survey Chief Edward Nelson pledged the federal government's commitment to "helping increase the amount of game." By the early 1920s, Nelson argued on behalf of the stock growers: "Inasmuch as they are paying grazing fees on the national forests they should have the protection of predatory-animals thereon at Federal expense."[244] From a game perspective, the results of PARC were disappointing. States had cut bag limits, enforced game laws, killed predators, and introduced or reintroduced game species, but the hunting was no better than it had been. In some cases it was worse. Ducks had been subject to federal regulation since 1918, but the 1931 duck season had to be cut from three and a half months to one month to preserve the diminished populations that remained. Bobwhite quail, then the most popular upland game bird, were so scarce that some legislatures reclassified the species as a songbird and ended its hunting altogether. Deer populations rose and fell in unpredictable ways. Starvation and disease struck seemingly secure populations.

There was policy—in the sense of agreed-upon actions—but it was not working. The most spectacular failures were in deer "management," and among these the fate of the herd in the Kaibab National Forest on the North Rim of the Grand Canyon stood out. In 1906 the federal government made it a game preserve. The herd grew rapidly until the mid-1920s, when the policy of protection seemed successful. Then animals began dying and the herd declined for a decade. More than any other single incident, the Kaibab experience forced a reevaluation of wildlife management theory and methods. And though there were no wolves and few other predators in the forest, the episode helped shape ideas about predation and predators.[245]

Milton P. Skinner, a former ranger and chief naturalist in Yellowstone, presented a paper at the organization's 1924 meeting calling into question Park Service predator control policies. Published a few years later in the *Roosevelt Wildlife Bulletin*, Skinner's critique highlighted "how unscientific" and

"careless" the federal approach to predatory animals in Rocky Mountain National Park had been. Skinner asserted that mammals required "predatory
enemies to keep them at the top-notch of efficiency." They listened. By the
early 1930s, the Park Service—being less affected, and therefore less influenced by livestock or hunting concerns than the Biological Survey, Forest
Service, or state natural resources departments—had banned steel traps and
poisoning.[246]

The Park Service also began to study the wildlife inhabiting its holdings.
In 1928 part-time ranger George Wright convinced his superiors to begin a
survey of wildlife. Three years later the Park Service appointed him field naturalist and in 1933 made him chief of the new Wildlife Division. The Wildlife
Division published Wright's faunal surveys as the first of its new series of
monographs: *Fauna of the National Parks*, which not only reported species
composition, but also advocated. In particular, it called for a new park policy
that all species should be preserved. Each species being "the embodied story
of natural forces which have been operative for millions of years and is therefore a priceless creation, a living embodiment of the past." Protection was
critically important for the larger predators. It was only in the parks that they
could "find their only sure haven . . . [and be] given opportunity to forget that
man is the implacable enemy of their kind, so that they lose fear and submit
to close scrutiny."[247] Further, he stated, "the park standpoint is quite different.
It has a special duty to protect the carnivorous forms which are blacklisted
everywhere else."[248] "To put it another way, we cannot stress the value of one
animal at the expense of another, for if we do our lopsided vision is reflected
in poor management which wrecks the whole organic wilderness." Unfortunately, the deaths of Wright and Yellowstone superintendent Roger Toll in an
automobile accident in February 1936 dealt a fatal blow to the Wildlife Division's emerging ecological vision. After the Wildlife Division was transferred
to the Biological Survey in 1939, then became absorbed with it into the Fish
and Wildlife Service in 1940, its influence dissipated.[249]

As the federal PARC program expanded, western mammalogists—led by
the Oklahoma-born Joseph Grinnell, director of the Museum of Vertebrate
Zoology at the University of California, Berkeley—moved to stop what they
perceived to be unnecessary killing. Grinnell was committed to the preservation of native wildlife and stubbornly opposed to the "absolute extermination
of any native vertebrate species whatsoever."[250] He became a mentor to some
of the most steadfast critics of the Survey's control operations.

Grinnell was adamant but not inflexible. He maintained: "It is perfectly
proper to reduce or destroy any species in a given neighborhood where sound
investigation shows it to be positively hurtful to the majority of interests."

Grinnell agreed with Muir: "The loss of a few chickens from hawks and a few calves from coyotes amounted to nothing compared with the toll levied on everything" by the burgeoning population of rabbits. Farmers in the state organized tremendous campaigns to wipe out the jackrabbits, "all without avail. The unthinking destruction of coyotes, hawks, and snakes backfired and exacted a still "penalty for interfering with the balance of Nature."[251] Grinnell pushed for a ban on livestock grazing in national forests. He also agreed with Ding Darling that science should take precedence in, and politics be removed from, the wildlife management decision-making process.[252]

Grinnell argued for the rights of predators in his September 1916 *Science* article coauthored with Tracy Storer.[253] With "occasional exceptions, according to season, place and circumstance," the authors called for the complete protection of predators.[254] The issues that Grinnell and Storer raised divided the American Society of Mammalogists. This society, formed in 1919 with C. Hart Merriam as its first president, included university-based scientists such as Grinnell, as well as a large number of Survey biologists, many hired by Merriam. Dissent between the commercial harm versus ethical rights of predatory mammals emerged openly and respectfully during a 1924 meeting at Harvard, where thirty-seven members participated in a symposium addressing the issue.[255]

Lee R. Dice—a former student of Grinnell's who was by then a professor at the University of Michigan and who would later oversee Adolph Murie's graduate work—argued for the intrinsic value of all mammals. Dice emphasized the importance of predators to the study of many aspects of biology, including anatomy and embryology, paleontology, biogeography, natural history and life history, as well as evolution. "With the predatory mammals eliminated it will become more difficult to explain the origin of many adaptive structures and habits in the remaining species."[256]

Joseph Dixon, from Grinnell's Museum of Vertebrate Zoology, presented data on the contents of over 2,500 stomachs. Mountain lions fed on deer, not livestock, while coyotes fed on carrion and rodents, not sheep. "In view of these and other similar findings, I believe that, in many instances where extermination is advocated, we are getting 'the cart before the horse.' I firmly believe that it would be best for all concerned to have a fair and impartial determination of the facts as a basis for a much-needed, sane, administration of the animal assets (or liabilities) in question."[257]

University of Chicago–trained Charles C. Adams, on the faculty of the powerful New York State College of Forestry, reiterated Dice's claims about the scientific value of predators, extolling their virtues as "one of the most powerful elements in preventing over-population by the herbivorous animals. . . .

FIGURE 48. Joseph Grinnell examining a mammal skull, 1930. Photo used with permission of the Museum of Vertebrate Zoology, University of California, Berkeley, image 8421.

The balance of nature depended on predatory animals' survival." Adams then proposed a moratorium on "all control measures within the parks, with some of the millions then devoted to predator control and boundaries diverted to scientific study and policy formation."[258]

A. Brazier Howell howled: "There is absolutely no excuse for a Government bureau, supposedly scientific, to have carried on this war of extermination for 14 years without a proper investigation of food habits. The drift away from its scientific mission, 'at the demand of the sheepmen minority, who have almost ruined the western mountains by overgrazing,' had drawn the Survey into a 'biologically unsound and exceedingly dangerous' policy."[259] Howell noted that the Survey had been transformed from a scientific to a political and economic entity led primarily by non-biologists. "The Biological Survey is potentially the most powerful single factor, for good or evil, that we have in this field," Howell stated. "For many years it was the respected and trusted public guardian of our fauna, but whether we will or no, conditions are now such that we are forced to label it not our federal wildlife warden, but the guardian of the sheep men and other powerful interests."[260]

The Survey fought back. A. K. Fisher wondered, "[J]ust how carefully all these men who are opposing this have been over the Western United States. . . . I am just as fond of wild animals as anyone but when they come up against man we have to do certain things for the benefit of man."[261] The criticism of the Survey took on a new tone, however, when it came from one of their own. In January 1931, Olaus Murie, then a PARC supervisor in Jackson, Wyoming, wrote a long letter to W. C. Henderson, assistant chief of the Survey, to "get things off my chest." Murie was having trouble with the Survey's handling of predator control. Instead of giving him the data he needed to form his own opinions, people were trying to convince him that the Survey was right. They had even offered him bribes—"the chance to hunt ducks without a license, etc." He reviewed the American Society of Mammalogists' complaints of the Survey's actions and justifications, and read the Survey's response. His conclusion: "To tell the truth I could not see that the queries of the Society had been adequately met."[262] The Survey, Murie pointed out, had killed predators in Alaska without benefit of scientific studies. It had relied on the "usual run of sourdough information . . . accepted at face value." Wolverines were being trapped, but, Murie wrote, "I have yet to meet anyone who knows anything about wolverine food habits in a quantitative way." Are we, he asked, "increasing the list of predators just because we are on the ground and in the business?" He knew his views were not popular, so he concluded by asking, "Am I a black sheep in the Bureau fold now?"[263]

Half a year later, Murie wrote to Howell, admitting he was "very fond of native mammals, amounting almost to a passion" and "would make considerable sacrifice for the joy of animal companionship and to insure other generations might have the same enjoyment and the same opportunity to study life through the medium of the lower animals." He thought the cougar was "nature's masterpiece in physical fitness," the wolf a "noble animal, with admirable cunning and strength." He wanted even the "so called injurious rodents around." He believed in control but only if the decision was made on a scientifically sound basis. We are "passing around an appalling amount of misinformation about the effects of predators on game. I have been awakening to the fact only in the last few years."[264]

In 1933 FDR became president, and his administration almost immediately reduced PARC funding.[265] Two years later, Roosevelt appointed Ding Darling head of the Survey, and Darling called PARC funding "turbulent and unsatisfactory."[266] Then, in 1936, Ira Gabrielson became Survey director and reversed the trend.[267] In 1939 alone, his agents killed 104,000 predators, including 93,000 coyotes.[268] Under Gabrielson, the body count soared, but the justification shifted: "The fact remains that . . . the Federal Government has an distinct obligation to assist in the prevention of undue depredations to livestock and game species ranging on and adjacent to public lands . . . because of its financial loans to stockmen."[269]

In May 1939, the Biological Survey moved from the Department of Agriculture into the Department of the Interior. A year later, it was combined with the Bureau of Fisheries and renamed the U.S. Fish and Wildlife Service, with Gabrielson as its first director. Criticisms notwithstanding, PARC continued to operate largely as it always had.[270] Administrators attended meetings with livestock growers' associations and state fish and game officials. The research lab in Denver continued to experiment on "new types of poisons." In Pocatello, Idaho, a plant manufactured and distributed strychnine-infused baits.[271]

In his 1941 book, *Wildlife Management*, Gabrielson wrote:

> It is always to be kept in mind that the destructive effect of the predation here discussed is not so much upon the total population of domestic animals as upon the economic welfare of human beings dependent on those populations. . . . The margin of profit, which must furnish that livelihood, is usually small and a very modest loss may wipe that out. In such circumstances the problem becomes not one merely of animal interrelationships [ecology] but of economics and human economic welfare as well. So long as it is necessary for man to tend his flocks in regions seriously infested by predators, some form of control will be exercised. Theorizing as to natural balance of animal populations in the world will not change this fact. The only questions are, how much control is necessary and how shall it be undertaken? . . . Human interests, primarily economic, will always be paramount, and when predation on domestic animals is involved it is useless to explain it from the purely biological [ecological] point of view.[272]

(Despite all this activity and justification, a curious thing happened. While government hunters and trappers were killing 50,000 to 100,000 coyotes a year, coyotes expanded their range from Mexico to the edge of the Arctic

Circle, from sea level to timberline, and from rural areas into neighborhoods.[273] Absent wolves, coyotes thrived. It is now well known that through the ecological principle of competitive exclusion, the elimination of wolves enabled coyotes [and coyote increases reduced red fox numbers]. The opposite also happens. More recently in northern Minnesota, when timber wolves began reestablishing old territories, coyote numbers diminished and red fox became more common.[274])

Paul Errington, responding to Gabrielson's sentiments, stated:

> I do not and never did maintain that we can always ignore economics in our philosophies concerning predators and predation. I am willing to reiterate as many times as need be that depredations of large wolves upon livestock can be serious, that I know that large wolves could not be tolerated in a part of the country where they could do so much damage as, say, in South Dakota. There may be economic aspects to predation even by the much less formidable coyotes that man may be entitled to do something about, though, in protecting his interests, I do not think that he is entitled to overdo his campaigning.[275]

Leopold also responded. First, he opposed what Gabrielson would not admit—that Survey agents applied poison: "Artificial poisoning [Leopold cites two California studies on the effects of thallium] is often an unsatisfactory remedy because of its cost and the danger to game, livestock, and beneficial wildlife."[276] Second, he resorted to the Survey's own science—the data provided by Herb Stoddard's *The Bobwhite Quail: Its Habits, Preservation, and Increase*. Leopold approached the subject without giving credit to the parent organization: "I do not know who first used science creatively as a tool to produce wild game crops in America. . . . The idea was doubtless conceived by someone long before it was first successfully applied by the Biological Survey [Stoddard] in Georgia."[277]

Unlike his more natural history–based predecessors, Stoddard counted and measured. He counted the number of quail coveys, nests, and eggs/nest (clutch sizes). He calculated nesting success, re-nesting attempts, and age composition. He assessed mortality among life-history stages (eggs, chicks, and adults). He calculated the effect each species of predator had on the population. He banded birds to trace ranges and distances traveled. He plotted covey territories against different cover and food types. And he assessed the effects of various management techniques such as prescribed burning.[278]

Other studies followed Stoddard's, funded in part by the Sporting Arms and Ammunition Manufacturers' Institute. In 1934 W. L. McAtee reported that the Survey's Division of Food Habits Research was supervising or cooperating with work on Hungarian partridge in Michigan, bobwhite quail in

Wisconsin, ruffed grouse in Minnesota and New York, scaled quail in New Mexico, and Gambel's quail in Arizona[279]—projects Stoddard and Leopold had selected.

In 1929 Leopold decided to duplicate Stoddard's work at the other end of the bobwhite's range, in south-central Wisconsin (the project McAtee referred to, above). He gave the assignment to Errington. Absent Stoddard's plantations and sportsmen, Leopold selected five square miles of farmland near Prairie du Sac, Wisconsin, twenty miles northwest of Madison.[280] Following Stoddard's lead, Errington examined available habitats—agricultural fields, pastures, and woodlots. Once a week he counted every covey from the time it formed in the fall until it dispersed in the spring. He took notes on food, cover, snow, snow crust, ice, wind, temperature, quail mortality, and cause of death. Errington's data showed an unexpected pattern. Survival rates were linked more to the area where coveys wintered than to predator abundance. For example, an increase in gray foxes during the winter of 1933–34 did not increase quail mortality. Land that had, in the past, supported quail through the winter did so again, despite fox abundance, and land that had not, absent predators, did not. He concluded, "Within ordinary limits, the kinds and numbers of native flesh-eaters may not be of much consequence in the winter survival of wild northern bobwhite populations."[281]

Errington understood that people leaped from the known *fact* of predation to the unknown *effect* of predation—from the knowledge that some animals ate others to the assumption that this controlled the population. This was not true. Quail did not live in fear of their enemies, and no single factor controlled population growth. In 1938 Stoddard and Errington coauthored a paper[282] emphasizing the variable impacts of predation across quail populations.[283]

These results undercut basic assumptions about predators and prey. In *Game Management* (1933), Leopold relied heavily on the work of both Stoddard (mentioned on eighty-five pages) and Errington (mentioned on thirty-nine pages) to argue against predator control as standard practice. Predator control might be necessary, he said, but it had to be shown to be effective in each case—one could not assume eliminating predators would improve game numbers.

Olaus and Adolph Murie complemented Stoddard's and Errington's work by studying the predators. As mentioned above, Olaus studied the coyotes of Jackson Hole for the Survey between 1927 and 1932.[284] Murie's results challenged the prevailing wisdom that coyotes were preying on elk, particularly calves. His analyses of coyote scat revealed little supplemented with carrion. To avoid being defunded, Survey administrators delayed publishing Murie's

Jackson Hole research until after passage of the federal Animal Damage Control Act in 1931.[285]

Olaus's half-brother Adolph, working for the Park Service, studied the food habits of coyotes in nearby Yellowstone between 1937 and 1939. Adolph found rodent remains comprised the majority of coyotes' diet, especially during the summer, while elk carrion made up a more significant proportion of their winter diet, but only about 16 percent overall. "Special emphasis was given to the task of determining the effect of coyote pressure on prey species," Murie wrote in his conclusion. "The facts show that in the case of elk this is negligible, and that no appreciable inroads on the populations of deer, antelope, and bighorn are taking place." Of more consequence, the report found that a modest amount of coyote predation on elk calves would do insignificant damage. Murie advised against "artificial control . . . under present conditions." Some Park Service officers wanted to fire him.[286] Murie proceeded, unafraid. In *Ecology of the Coyote in the Yellowstone*, he wrote: "Since the advent of the modern conception of wildlife management a new attitude toward the question of predation is growing. One of its precepts is that control of potentially harmful or suspected species of birds and mammals should await precise data based on research."[287]

Errington also questioned the necessity and methods of predator control, emphasizing instead the ecological balance that wildlife managers should strive to understand and maintain. In a 1937 article, he pointed out that predators "are only animals and their lives are circumscribed by natural limitations just as are the lives of animals which are not usually thought of as being predators." He criticized conservation as then practiced for defaulting to "persecution of predators" without seriously considering worthwhile alternatives or contemplating the importance of predation in terms of a given area's carrying capacity for prey species. Ten years later, in an influential essay published in the *Journal of Wildlife Management*, Errington criticized hunters in particular for their insistence on "vermin" elimination. "Suppression of predators or competitors does not necessarily mean benefit to the game or more game to shoot; it is not a panacea and on the whole—not entirely—shows less promise with investigation."[288]

Survey workers saw game management as a new, less controversial mission for the agency. W. L. McAtee was the most prominent advocate. McAtee had started with the Survey in 1903,[289] analyzing the food habits of birds from stomach contents, and had risen to head the Division of Food Habits Research. He was displeased with the Survey's growing commitment to predator control and concomitant loss of scientific credibility. When McAtee lost his position in 1934 (for which he blamed Clarence Cottam but should have blamed Ding Darling), he created a new niche for himself as the Survey's go-to person for information on wildlife research. McAtee helped form the Wildlife Society in 1936 and two years later became the first editor of its publication, the *Journal of Wildlife Management*.

The Survey had always been responsible for the administration of federal wildlife policies, and FDR's New Deal programs gave them more work. In 1935 the agency began overseeing Darling's cooperative wildlife research units. In 1938, after Congress passed the Pittman-Robertson Act, which provided federal funds to state departments of natural resources for game research, the Survey set guidelines and enforced standards.[290]

While the National Park Service played a key role in protecting predators, even this agency could not completely escape the biases of the times. In 1926 Park Service director Stephen Mather stated it was not park policy to exterminate any native animal, although he did endorse predator control to protect "weaker" species. Two years later, park superintendents formally condemned predator control. In 1931 Mather's successor, Horace Albright, banned poison in the parks. The Park Service, he declared, would "give total protection to all animal life."[291] George Wright's Wildlife Division funded and published Adolph Murie's work on coyotes and wolves. It also supported the drive to make Isle Royale a national park preserved as a wilderness area. When wolves colonized the island in the early 1950s, the Park Service welcomed the opportunity, and in 1957 it contracted with Purdue's Durward Allen to study the relationships between wolves and moose.

Not all Park Service affiliates were onboard. In 1937 and 1938, retired Park Service director Albright reversed himself and condemned Adolph Murie's research on Yellowstone's coyotes. In typical administrator-think, Albright felt predators should be reduced or eliminated because visitors never saw them. Ranchers surrounding national parks were equally annoyed, particularly those affected by the grizzlies sheltered in Yellowstone.[292]

In 1947 the powerful compound 1080 (sodium fluoroacetate) became available.[293] At the same time, western livestock interests began facing competition for access to public lands from recreationists.[294] Experiencing the outdoors and seeing wildlife, not shooting it, drew people to state and federal lands. "Sportsmen have their privileges, and no one wishes to take away that which is rightfully theirs," National Park Service biologist Victor H. Cahalane wrote in 1946. "But consideration must also be given to the desires of the non-hunting public. Those persons who are interested in all animal life should have an equal opportunity to see wildlife in a natural setting and natural distribution."[295]

A reformed Sig Olson chimed in: "With the fast-growing appreciation of the true meaning of wilderness, we are beginning to question the idea of the total elimination of predators, realizing that, after all, lions, wolves, and coyotes may be an exceedingly vital part of a primitive community, a part which once removed would disturb the delicate ecological adjustment of dependent types and take from a country a charm and uniqueness which is irreplaceable." Herbivores unconditioned to predatory rivals lost their "natural alertness," and wilderness thus deprived of its balance became something akin to "a cultivated estate."[296] Also, Leopold felt the key to managing wild animals centered on improving a species' environment, rather than just artificially adding animals to the landscape.[297]

As executive director of the Wilderness Society, Olaus Murie felt that large wilderness areas were the key to saving predators. To preserve "our big, interesting creatures, such as the wolf, grizzly bear, and mountain lion," he wrote several years later, "we have to deal principally with human attitudes" and offer "ecological facts" in the hope of diverting official wildlife management policies from their established channels. "Scattering deadly poison over our landscape, with its resultant destruction of wolves and other animals, to the accompaniment of lurid hate-propaganda, completely avoids the issue and offers no permanent policy for the welfare of our game herds."[298]

The growing outcry by wildlife advocates over predator control influenced Michigan congressman John D. Dingell to introduce legislation in 1963 to strip PARC of appropriations and personnel, and restrict it to advisory duties. This compelled President Kennedy's secretary of the interior Stewart Udall to task an advisory board (the Leopold Committee) to survey wildlife management policies with reviewing the whole Predator Animal and Rodent Control program.[299] The Leopold Committee had clout, including Aldo's son, Starker, Cottam, and a reformed Gabrielson, who had left the Fish and Wildlife Service in 1946 to first head the Wildlife Management Institute, then the World Wildlife Fund. By the early 1960s, "Gabe" had evolved into a defender of predators.[300]

In early 1966, Congressman Dingell convened hearings on predator control in his Fisheries and Wildlife Subcommittee. Dingell warned the federal government to take reform of predator control practices seriously to, as he stated to Stanley Cain at the outset of the latter's testimony, "restore some sanity to what in my opinion has been a rather excessive program to destroy these mammals."[301] "Although the attitude at the three or four highest levels of administration in the Department of the Interior is enlightened and therefore offers the opportunity for needed change in policy, the 700 to 900 employees in the field are not about to change. . . . The woolgrowers are not about to foster a change to a different system."[302]

In 1970 a National Academy of Sciences report expanded upon previous inquiries about the sheep industry's grip on policymaking. "It is illogical for the government to support a private commercial endeavor where the cost of protection exceeds the value of that which is protected and where the protection program conflicts with the other publically owned natural resources."[303]

Despite these arguments, it would take the discovery of un-nature in nature by Boy Scouts (similar to the malformed frog phenomenon twenty-five years later[304]) to spark a public outcry that would force society to rethink itself. In May 1971, the backlash after a Scout troop found thallium-poisoned golden and bald eagles near Casper, Wyoming, did more than any scientist ever could have to galvanize public demands to end government-sponsored predator control.[305]

Two months prior to this, in March 1971, Defenders of Wildlife and the Sierra Club sued the Department of the Interior for violating the National Environmental Policy Act (NEPA), the Endangered Species Conservation Act of 1966, the Animal Damage Control Act of 1931, and the Migratory Bird Treaty Act of 1918. The plaintiffs sought a preliminary injunction to halt the predator control program immediately, together with a permanent restraining order. A few weeks later, the Humane Society sued as well. The conservationists contended that because the Fish and Wildlife Service had not filed an Environmental Impact Statement, predator control was in violation of NEPA regulations. Furthermore, the 1931 law did not list several specific—and now endangered—species as targets for control, and predator and rodent poisoning had indeed decimated these species (California condor, black-footed ferret, San Joaquin kit fox, and Utah prairie dog). The Division of Wildlife Services fought the suit, and the preliminary injunction request was denied in October 1971, after which the plaintiffs and the Department of the Interior cut a deal. The conservationists agreed to drop the suit in exchange for an end to poisoning on federal lands by February 1972.[306]

Is there any more unlikely environmental hero than Richard Nixon? In 1972 Nixon proclaimed to Congress: "This is the environmental awakening. It marks a new sensitivity of the American spirit and a new maturity of American public life."[307] "I am today issuing an Executive Order [11643] barring the use of poisons for predator control on all public lands. I also propose legislation to shift the emphasis of the current direct Federal predator control program to one of research and technical and financial assistance to the States to help them control predators by means other than poisons."[308] Also in 1972, the Federal Insecticide, Fungicide, and Rodenticide Act strengthened previous

(1910 and 1947) legislation,[309] and the Federal Environmental Pesticide Control Act was passed.[310]

In 1962 the Bureau of Sport Fisheries and Wildlife's Committee on Endangered Species in the Division of Wildlife Research was established and charged with listing and determining measures to protect endangered species. The committee produced the first list of sixty-three endangered species in 1964. Congress responded to the extinction crisis as well, first with the Endangered Species Preservation Act of 1966, which, among other provisions, authorized limited purchases and management of habitats for the National Wildlife Refuge System to protect native wildlife listed as endangered. Three years later, the Endangered Species Conservation Act of 1969 strengthened the authority of the interior secretary to list foreign species and control their passage into and out of the United States, added invertebrates as a category of animal subject to listing, and increased funding for establishing refuges. Nixon felt that the 1969 law had failed to "provide the kind of management tools needed to act early enough to save a vanishing species." His proposed legislation that would "make the taking of endangered species a Federal offense for the first time, and would permit protective measures to be undertaken before a species is so depleted that regeneration is difficult or impossible."[311]

The Endangered Species Act passed by a vote of 92–0 in the Senate, 390–12 in the House. Nixon signed it into law on December 28, 1973. The "most radical piece of environmental legislation ever passed in the United States" barely set off a ripple.[312] Notable milestones: the timber wolf was ESA listed in 1967, and the Northern Rocky Mountain Wolf Recovery Plan was published in 1975.[313] The Mexican Wolf Recovery Team was formed in 1979, the same year the Grizzly Bear Recovery Plan began.[314]

Some numbers. For the cooperative predator control system from its inception in 1915 until the end of the 1939 fiscal year, a total of 1,174,084 predatory animals were destroyed. Of the total, roughly 88 percent were coyotes (1,031,951), followed by 110,642 bobcats and lynxes, 21,410 wolves (more than 17,000 being red wolves in Arkansas, Oklahoma, and Texas), 5,128 bears, and 4,953 mountain lions.[315]

The advances that Nixon championed in the early 1970s were partially or wholly reversed by his successors. In 1975 President Gerald Ford partially rescinded the ban on the use of poisons in predator control to allowed the use of M-44s containing sodium cyanide. And to no one's surprise, President Ronald Reagan revoked President Nixon's 1972 executive order, which again permitted the use of poisons on public land to control predators. Further, Reagan's interior secretary, James Watt, reinstituted denning—removing young animals from their dens by either "smoking, burning, or vacuuming" them out, then "burning, shooting, or clubbing them to death."[316] Therein lies the problem with executive orders.[317]

The U.S. Forest Service was born the same year the Biological Survey began its Predator Animal and Rodent Control program, and it found itself facing similar conservation/preservation conflicts. According to former U.S. Forest Service's deputy director Jim Furnish, the agency's history encompasses three periods. The first was Pinchot's custodial period, which lasted until just after the Second World War—into the mid- to late 1940s.[318] At this time, the nation's lumber was coming from private timbering companies clear-cutting their holdings.[319] Federal foresters such as Aldo Leopold did little tree cutting; instead they cruised and mapped the nation's forests and, with the help of work relief programs such as the CCC, built an infrastructure of access roads, trails, and telephone lines. In 1915 the Term Permit Act opened the way for a massive development of summer homes, camping areas, resorts, and hotels in national forests.[320] Furthermore, a series of Federal-Aid Highway Acts, beginning in 1916, provided $75 million over the next decade for road-building projects. Of this money, the Forest Service received $1 million per year. Additional legislation in 1919 and 1921 increased funding and distinguished between two types of roads: "forest development roads," designed specifically for administrative purposes such as fire suppression, and "forest highways," designed to augment state and county road systems and connect local communities. Road construction was the main reason visitors to national forests quadrupled between 1917 and 1924.[321] While foresters continued to cling to forestry, increased public use forced them to adopt unfamiliar recreational priorities.[322] So much so that in 1918, the Forest Service began hiring landscape architects.

Arthur Carhart, who had received his degree from Iowa State in 1916, was among the first of these landscape architects. He was assigned to District Two, which then comprised the Rocky Mountain and northern plains states, including Minnesota's Superior National Forest.[323] Carhart's plan was to develop national forests dedicated to recreation, free of commercialization, with public access to all. Carhart envisioned national forests with immense recreational spaces, with roads and trails, campgrounds and cabins, even resorts, and suggested a trolley form of public transportation to move people through the forest.[324] In 1919 Carhart visited Trappers Lake in Colorado's White River National Forest to plat sites for hundreds of summer homes. After some thought, he proposed no summer homes or hotels be developed along the lake's shoreline, but instead built no closer than half a mile from the lake, with lake access along trails.[325]

In December 1919, Leopold and Carhart spent a day discussing Trappers Lake and how such community planning might be applied elsewhere. Carhart followed up and offered that the Forest Service was ignoring its public duty by refusing to get engaged in recreational development.[326] Leopold was uneasy about the types of development that Carhart proposed. Aggressive development reduced the options of people seeking wilderness experiences, and Leopold did not want the Forest Service to get heavily involved in this form of recreation.[327]

A year and a half later, Leopold published "The Wilderness and Its Place in Forest Recreational Policy."[328] Leopold wanted preservation *from* certain forms of recreational development; particularly road building and term permit facilities.[329] He felt the Forest Service had caved to those seeking modern recreational access.[330] William B. Greeley, head of the Forest Service from 1920 to 1928, agreed and responded by proposing a four-stage process of wilderness identification within national forests, as follows: (1) district foresters identifying roadless areas offering unique opportunities for wilderness recreation; (2) these areas would then be protected from road building and term permit development; (3) in these areas, wilderness recreation would be the dominant use; and (4) these areas could be opened in the future if the nation needed their "economic resources."[331] This process culminated in Regulation L 20, enacted in 1929—the first formal Forest Service action designed to preserve wilderness. In a nod to the Ecological Society of America's Committee on the Preservation of Natural Conditions, L-20 further suggested that foresters set aside "research reserves" for scientific study.[332] While an advance on previous policies, L-20 decisions were made at the district level and were therefore applied unevenly.[333]

On September 19, 1939, FDR's secretary of agriculture, Henry Wallace, improved L-20 protections by issuing three U Regulations. Regulation U-1 read: "Upon recommendation of the Chief, Forest Service, national forest lands in single tracts of not less than 100,000 acres may be designated by the Secretary [of Agriculture] as 'wilderness areas,' within which there shall be no roads or other provisions for motorized transportation, no commercial timber cutting, and no occupancy under special use permit for hotels, stores, resorts, summer homes, organization camps, hunting and fishing lodges, or similar uses."[334]

Regulation U-2 set up a similar formula for "wild areas"—tracts of land between 5,000 and 100,000 acres. Besides the size of the area protected, the main difference between U-1 and U-2 was that all U-2 decisions rested with the chief of the Forest Service alone. The final level of protection, Regulation

U-3, provided for the creation of "roadless areas" of recreational value. These areas could be any size; areas of fewer than 100,000 acres could be classified by the chief of the Forest Service while areas over 100,000 acres required classification by the secretary of agriculture. Unlike "wilderness" and "wild" areas, timber cutting and other economic activities were permitted in "roadless areas."[335]

Which gets us back to Jim Furnish. The Forest Service's second period, the utilitarian era, ran from the postwar late 1940s through the early 1990s. It was a time of aggressive timber harvesting with forest management plans designed to meet high postwar demand for lumber. The Forest Service made national forest timber available to private companies, and companies arose that owned only sawmills.[336] In exchange for this milling capacity, the Forest Service supplied the timber.[337] The Forest Service's mission became harvesting national forests to fuel the expanding economy. It built roads, developed nurseries to provide the millions of seedlings necessary for planting clearcuts, and pioneered its Smokejumper program to get personnel on fires soon after they were spotted—keeping small fires small—so valuable timber would not burn.[338]

It wasn't just logging. The development of America's iconic ski areas following World War II was the idea of veterans of the Tenth Mountain Division, who understood that American terrain equaling Europe's great alpine ski resorts was designated as national forest. After climbing to the top of Vail Mountain in 1957, Pete Seibert and his group asked the Forest Service if federal lands could be developed for downhill skiing. The agency, in the belief that ski-area development was in the public interest, said yes. As a result, Vail, Aspen, Breckenridge, Winter Park, Steamboat, and other areas were developed.[339]

Anyone with a sense of history could have predicted the fallout from clear-cutting national forests. Back in 1924, Bob Marshall wrote "Recreational Limits to Silviculture in the Adirondacks." He was responding to the old forestry argument that timber harvest and outdoor recreation could coexist.[340] Marshall observed that because clear-cutting left great swaths of forest either cleared or worked over, these regions were aesthetically undesirable. Even if virgin stands were preserved in the name of recreation, the resulting fragmented landscape would disappoint those who sought a wilderness experience. Marshall was not objecting to silviculture in the broadest sense, but instead to the intensive silvicultural practices most foresters brought to regions becoming recreationally important.[341]

Furnish's utilitarian period had a good run before slamming headfirst into Judge William L. Dwyer, who, on March 7, 1991, ruled that the Forest

Service must ensure viable populations of northern spotted owls under the National Forest Management Act (NFMA).[342] (The Forest Service had argued, unsuccessfully, that because spotted owls were listed under the Endangered Species Act, the U.S. Fish and Wildlife Service was responsible for their recovery.) Because northern spotted owls required large, unbroken forest tracts, Dwyer's ruling forced the Forest Service to consider exactly what its harvesting practices and intensities were doing to the ecology of the nation's forests and grasslands—the regions it was charged by the citizens of this country to protect.

Furnish defined the third, current period in the Forest Service's history as ecosystem management, which arose in direct response to Dwyer's ruling. Furnish argued that because the Forest Service had historically paid little attention to ecological relationships—ecology was never a component of its original mission—it had struggled, and continues to struggle, to meet this responsibility.[343] Regulation L-20, which was established in 1929, and Regulations U-1, U-2, and U-3, established in 1939, were acknowledgments by the Forest Service that it recognized the Ecological Society's Committee on the Preservation of Natural Conditions' wish to preserve representative landscapes as baselines for ecological studies.

Furnish noted that a social contract exists when a federal agency stewards a resource in trust for a public constituency: that the Forest Service—following the spotted owl decision—should have recognized they had violated this social contract and reconsidered their responsibilities to reestablish their credibility in the eyes of a skeptical public. Instead, the Forest Service, much like the Biological Survey did with its Predator Animal and Rodent Control programs, battled for the status quo.[344]

By the 1960s, the public was angry over Forest Service management plans, held large demonstrations opposing logging, and filed lawsuit after lawsuit to protect threatened and endangered species.[345] Politicians responded. First with the Clean Air Act, then with the Wilderness Act, the 1970 National Environmental Policy Act,[346] the 1972 Clean Water Act,[347] the 1973 Endangered Species Act,[348] and the 1976 National Forest Management Act.[349]

Disagreement with Forest Service policies grew. Environmental organizations such as Save the Redwoods and the Sierra Club established grassroots support, then grew rapidly in size and influence. They began hiring lawyers and used the new environmental laws to defend their interests.[350] For example, in the early 1970s, the Izaak Walton League sued the Forest Service on the grounds that clear-cutting West Virginia's Monongahela National Forest violated the requirement of the Organic Act of 1897 that trees be individually marked for selective logging.[351] It won. As Furnish points out, while a Forest

Service supervisor can bully a fisheries biologist, they cannot intimidate a federal judge.[352] With forest management, especially old-growth forest management, society had changed course, shifting away from utilitarian, economic drivers toward more sustainable, ecological metrics,[353] or, in other words, shifting from conservation in Pinchot's sense to preservation in Muir's sense.

As with the Biological Survey, the Forest Service had its internal critics.[354] Dissenters' views were reinforced by new employees trained in ecology, hydrology, and wildlife management, hired to provide the expertise necessary for the Forest Service to comply with new federal legislation. They formed the nucleus of a new group—the Forest Service Employees for Environmental Ethics (FSEEE)—and began calling out hard-line old-guard foresters out of compliance.[355] In 1993, two years after the northern spotted owl ruling, wildlife biologist Jack Ward Thomas became chief of the Forest Service.[356] His first action was to issue a six-point directive that included mandates to tell the truth and to obey the law. Implicit in Thomas's action was that Forest Service employees had lied and broken the law.[357]

Following Thomas's tenure, another biologist, the ichthyologist Michael Dombeck, became Forest Service chief. Dombeck, appointed during the second Clinton administration, managed to push an executive ruling permanently protecting roadless areas within the national forests. Smaller roadless areas (designated U-3 in 1939) had historically been regarded by the Forest Service as simply the next place to look to cut trees. Despite this attitude, in 2000, 58 million acres of roadless areas—about 2 percent of the land area of the United States—remained. Congress had previously created about 36 million acres of wilderness, making around 94 million acres of wilderness. This meant that as the millennium turned over, close to half of the 192 million acres of national forest remained undeveloped.[358] Furnish notes that at the time, wilderness areas were predominantly high-elevation lands, while roadless areas tended to occupy mid-elevation lands more vulnerable to cutting.[359] Alaska's two national forests, the Tongass and Chugach, contained about 12 million of the 58 million acres of roadless areas, more than 20 percent. The Tongass was the crown jewel of all national forests and became a political hotbed.[360] Dombeck felt that "Alaska's roadless lands are the best we have. We need to protect the best."[361]

The executive order establishing the roadless areas had to be, and was, completed before George W. Bush was inaugurated in January 2001. Although the Bush administration immediately negated the ruling, in 2012 the Tenth Circuit Court of Appeals reversed Bush's action, the Supreme Court let the lower court's decision stand, and the ruling became law.[362] This is the great thing about executive orders.

Today's foresters assume a more ecological way of considering nature.[363] For example, today's goal in the Black Hills is to create a healthy, more natural forest that is more resistant to crown fires and insect infestation. Foresters there ask themselves how they can design an ecologically sound forest, not just a tree farm.[364] The battle is not over, especially in remote areas such as the Tongass and elsewhere in Alaska,[365] and there will likely be setbacks for a long time to come, but the trajectory is resistant to change.[366]

What It Meant

There can be no doubt that despite much resistance and many reversals, during the twentieth century, society shifted its environmental ethic from conservation to preservation, from use to no use. National forests, national parks, national wildlife refuges, and BLM lands all contributed wilderness areas. Much of the hard work in defining and describing these areas was done by field biologists. But as field biologists multiplied, they became more and more embedded in organizational infrastructures, and as a result of this assimilation became less visible. As I write this, the National Park Service is celebrating its centennial. Who gets mentioned most often? Stephen Mather, the first director. I wonder how many holdings Mather personally scouted and proposed for inclusion.

We can divide the old wildlife biologists and foresters into three camps: those, such as Edward W. Nelson, who never shifted from their utilitarian roots; those—such as Aldo Leopold, Sigurd Olson, Paul Errington, Ira Gabrielson, and Jim Furnish—who switched mid-career from practicing utilitarian conservation to embracing wilderness-level preservation; and those, such as Olaus and Adolph Murie and Bob Marshall, who were always preservationists at heart. As much as anything, this may have been a generational phenomenon, with the oldest biologists being the least willing to change. But I cannot help but feel that this willingness to change also came down to quality of character.

The Restoration Biologists

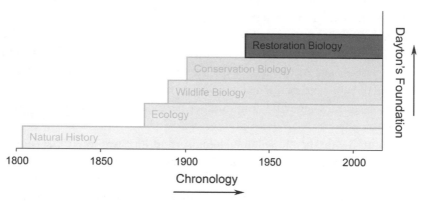

FIGURE E.

Once ecosystems have been lost or degraded (including, in the big picture, losing their top predators or having their canopy structure obliterated by logging), it falls to the restoration biologists to bring them back. These people, knowledgeable and dedicated, work by first determining which critical species composed the original landscape and which ecological processes they provided. They also consider factors such as soil composition, hydrologic processes, local ecotypes, and perhaps special considerations such as endangered species requirements. Establishing these landscapes can be costly and time intensive. Once they have been recovered or reborn, they must be celebrated and enjoyed; they should also be monitored.

As important and necessary as restorations are, even big restorations can lack critical biological phenomena such as migrating animals. While birds, and to some extent migratory butterflies, can fly over inhospitable landscapes and find and use these oases of habitats, more earthbound species must discover these newly formed sites using, literally, more pedestrian means. How much land must be restored to accommodate the historic seasonal migrations of bison, elk, and caribou? And how much more land is required to hold the wanderings of grizzly bears, wolves, and cougars? Even the lowly amphibians, which cannot flee far or fast in the face of habitat destruction, must have the space to breed in wetlands, feed in uplands, and overwinter in lakes, as well as the ability to hop or crawl safely among these ecosystems.

For mobile species to persist, restoration biologists must create dispersal and migratory corridors—in a tinker-toy metaphor, the sticks must connect the spools. Dave Foreman's Wildlands Project (now the Wildlands Network[1]) has been a leader in calling attention to this problem and providing solutions.

All the King's Horses and All the King's Men

As Andre F. Clewell and James Aronson described (italics theirs):

> Ecological restoration is the process of assisting the recovery of an ecosystem that has been degraded, damaged, or destroyed. From an ecological perspective, it is an intentional activity that reinitiates ecological *processes* that were interrupted when an *ecosystem* was impaired. From a conservation perspective, it recovers biodiversity in the face of an unprecedented, human-mediated extinction crisis. From a socioeconomic perspective, *ecological restoration* recovers *ecosystem services* from which people benefit. . . . While globally cumulative, ecological restoration is necessarily a local endeavor.[2]

Susan M. Galatowitsch continued the thread to include content and integrity:

> Restored ecosystems should: 1) contain a characteristic assemblage of species that occurred historically or currently occur within a reference ecosystem; 2) consist of local ecotypes of native species to the greatest extent possible; 3) contain all components of normally functioning ecological groups; 4) be integrated into the larger landscape; 5) be resilient to stressors; and 6) be self-sustaining.[3]

The science of ecological restoration began with a prairie reconstruction.[4] When Aldo Leopold spoke at the University of Wisconsin's Arboretum dedication, on June 17, 1934, he observed that while older arboretums were "collections of trees," modern arboretums were being arranged around natural plant associations including trees, sub-canopy shrubs, and native ground cover. Leopold's idea for the Wisconsin Arboretum was "to re-construct . . . a sample of original Wisconsin—a sample of what Dane County looked like when our ancestors arrived here during the 1840s." It would not be done for "amusement," but instead for research. Leopold envisioned an oak savanna growing in a field of prairie grasses and forbs to "serve as a bench mark, a starting point, in the long and laborious job of building a permanent and mutually beneficial relationship between civilized men and a civilized landscape."[5] Leopold considered natural preservation and restoration for aesthetic and scientific purposes a better use of land than recreational park systems.[6]

Leopold's colleague Norman C. Fassett is usually credited with the idea of restoring tallgrass prairie. Fassett was the curator of the University of Wisconsin's Herbarium and was the first to sow prairie seeds at the Arboretum, which was the first native grassland restoration attempted. A Special Committee on Arboretum Planning was formed in December 1933, composed of Fassett, Leopold, and three others.[7] The following month, the committee proposed several research projects including studies on invasive carp to be supervised by Chancey Juday and an experimental prairie grass planting to be supervised by Fassett.[8]

When Fassett began spreading prairie seed, "little was known about planting procedures," and he soon discovered the highest survival occurred when prairie sod was transplanted and "prairie hay"—stems and seed heads—was scattered in bare areas.[9] Not long after, Ted Sperry's dissertation research on prairie root systems was published in the journal *Ecology*,[10] and the Special Committee on Arboretum Planning knew they had identified the man to carry out their restoration project.

Theodore M. Sperry was born in Toronto, Ontario, on February 20, 1907. Soon after, his family moved to a home near Indianapolis, and Sperry grew up exploring the countryside. He received his BS from Butler University, then went to the University of Illinois for both his master's and PhD degrees in botany. His advisor was Arthur Vestal, who received his PhD at the University of Chicago working with Henry Cowles. Despite this Chicago connection, Sperry's work was deeply influenced by Frederic Clements and John Weaver, out of Nebraska.[11]

After a long, politically motivated hiring period, Sperry took over the Arboretum restoration, with the assistance of the Civilian Conservation Corps (CCC) unit stationed at Camp Madison. Leopold wrote a letter to Sperry prior to his hiring: "We now have a CCC camp. . . . On the area is an open flat Wisconsin prairie, together with its accompanying 'oak openings.' . . . You can appreciate the difficulty and variety of the technical problems involved."[12]

Sperry began in February 1936.[13] Because there was no literature, his early efforts were guesswork and intuition. According to Franklin Court, "Not even

Fassett knew exactly how to turn a plowed field with 'corn stubble, quack grass and rag weed' into a native Wisconsin tall grass prairie."[14] Sperry found prairie remnants north of Madison, near Prairie du Sac, and up in Sauk County, in an area Leopold had explored when looking for his weekend getaway. Sperry and his crew could find and collect this prairie sod in the Wisconsin River flood-plain, but he needed a plan. The question was "Were we going to make a care-fully organized garden or just going to try to cover an area?" Sperry decided to spread the sod thin, so "it would look like we were getting the job done."[15]

Sperry described their technique:

> We took spades and mattocks. We dug a hole. We got rid of as much quack grass as possible and we put the sods right in there. . . . We got out the quack grass, loosened the soil and put the seeds a quarter of an inch below the soil.[16]

Sperry did not want a monoculture. He tried to mix species in patches and assumed that after they took root, shed seed would eventually fill in bare areas. Sperry later said there were no set goals for abundance and distribu-tion; that he "never knew specifically what he was supposed to be doing."[17] In 1939 he started a prescribed burning regime to reduce non-native species and tree seedlings. Between 1936 and 1941, "Sperry's weed patch" "took."[18] By 1939 Sperry's prairie had a reputation, and Robert Mann, the superintendent of maintenance of Cook County, Illinois (Chicago), asked Leopold if he could borrow Sperry to develop a prairie for his Forest Preserve District. In 1962 Sperry's 60-acre restoration was named Curtis Prairie, after the famous Wis-consin botanist John T. Curtis, following his untimely death.[19]

In October 1970, the *New York Times* published a long article on national efforts, begun in Wisconsin and Illinois, to preserve and restore "vanishing grasslands," which stated that along "with Willa Cather [in *My Antonia*], they feel 'the grass was the country,' and that its roots ran deep and shaped the heartland of a nation."[20]

The University of Wisconsin Arboretum restoration efforts did not stop with prairie. The Arboretum Commission and the Wisconsin Conservation Department (today the Department of Natural Resources) were concerned with the damage that European carp had done to the emergent aquatic vegeta-tion (sedges, rushes, and cattails) along the Lake Wingra shoreline, and they proposed an eradication program.[21] On November 15 and 18, 1936, these groups ran a 5,000-foot (1.5-kilometer) seine through Lake Wingra and removed 71,550 pounds (32,454 kilograms) of fish, 58 percent of which were carp.[22]

The Arboretum also proposed a woodlot restoration, to be conducted throughout the state. However, woodland restoration techniques were "vir-tually unknown, because almost all forestry research programs had been directed toward pine, spruce, and northern hardwoods, rather than oak-hickory woodlots." Potential projects included measuring success rates based on seeding versus nursery transplants.[23]

Herb Stoddard had his own ideas about restoring savanna, in particular longleaf pine stands, which included prescribed burning.[24] Both longleaf pine and prairie ecosystems had been managed for millennia by Native Americans using fire, making these "semicultural" ecosystems.[25] Stoddard detailed his reasoning:

> The southeastern pine forests are true fire types; the animal and vegetable life of the pine belt had adjusted through the ages to periodic fires—in the case of the longleaf, perhaps as often as every two to three years. The ecological upset, caused largely by the efforts of well-meaning but uninformed professional foresters, had brought about serious conditions, among them a decline in quail. . . . [I]n the early years of our investigations it was considered heresy by most professional foresters to recommend burning in the pinelands.[26]

Restoration biology may be the epitome of "think locally, act globally." Guided by knowledge of ecological processes and a notion of what has worked in the past, restoration ecologists must have a deep knowledge of local ecosystems and be willing to think creatively about the means necessary to accomplish their management ends. For restorations to be done correctly, field biologists must be present at every step. It is a new field and is still developing its legacy, but its potential is being considered. A recent headline addressing the potential for humans to colonize other planets reads: "How we can build another Earth."[27]

Indiana Jones Revisited

How could the midwesterners Roy Chapman Andrews and Gordon Mac-Creagh not have inspired the Indiana Jones character? Jones is, of course, an amalgamation of influences (only George Lucas and Stephen Spielberg really know their inspirations, and they're not telling) but certain traits of both Andrews and MacCreagh seem too similar to Jones's to be pure coincidence.

Roy Chapman Andrews was born in Beloit, Wisconsin, on January 26, 1884. He claims, "The greatest event of my early life was when, on my ninth birthday, Father gave me a little single-barrel shotgun."[28] His favorite childhood book was *Robinson Crusoe*, and from as long as he could remember, he wanted to be an explorer. Andrews claimed that by using William Temple Hornaday's "*Taxidermy and Home Decoration*," he taught himself taxidermy, but this must be a humorous mis-memory. In 1891 Hornaday published *Taxi-*

dermy and Zoölogical Collecting, which rolls off the tongue about the same way as taxidermy and home decoration, but Hornaday never published a book with this title, and his professional interests never included interior design.

Andrews attended Beloit College (the "Yale of the West," which Stephen Forbes had attended almost thirty years earlier, and where Ding Darling had recently graduated). While in college, Andrews nearly drowned in a canoe accident that killed his friend Monty White. Badly shaken, Andrews lost his hair and developed a nervous affliction, which he retained for the rest of his life.[29] While he was a senior at Beloit, Edmund Otis Hovey, curator of geology at the American Museum of Natural History, visited Beloit to give a lecture on the eruption of Mount Pelée. Afterward, Andrews spoke with Hovey and through him established a correspondence with Hermon Bumpus, the director of the American Museum of Natural History. Andrews then went to New York and convinced Bumpus to hire him to mop the great museum's floors and work other odd jobs.

After a few months, Andrews helped construct the museum's first suspended whale exhibit. He was then asked to salvage a whale killed off Long Island, which led to his study of whale breeding off the coast of British Columbia. He went on to collect beluga whales in the St. Lawrence, which in turn led to a whale expedition in the Far East, where he rediscovered gray whales, which had been extirpated along the American coast and presumed extinct.

During his first trip to Asia, Andrews was based on the *Albatross*, Spencer Baird's old research vessel out of Woods Hole. In her day, the *Albatross* was the most famous oceanographic vessel on Earth—like Jacques Cousteau's *Calypso* or any ship that Robert Ballard happens to be on—doing hydrographic work as well as deep-sea seining and dredging.[30] As Andrews described:

> She was a beautiful ship, built like a yacht. With a wide afterdeck where the officers slept on camp beds when the night was hot. . . . She was a "bastard" ship according to the Navy, for she was owned and controlled by the U.S. Bureau of Fisheries but manned by the Navy. None of the officers cared for the duty because they felt that, while it was pleasant enough, it did not advance them in their profession. It wasn't a "happy ship." Most of the scientific staff as well as the officers had been aboard too long, and friction had developed to such an extent that several were not on speaking terms with the others.[31]

Tensions aside, and reminiscent of Ferdinand Hayden forty-five years previously, the scientists and crew of the *Albatross* played a baseball game against a team from the Pacific fleet admiral's flagship, the *Rainbow*. With Andrews playing catcher, the *Albatross*'s team won.

On Andrews's second Asiatic trip, he collected specimens in the northern Korean larch forests. He was so far behind schedule, he was thought to be dead. When he returned to Seoul, he had the third-person pleasure of reading his own obituary.[32]

In 1917, just after the United States declared war on Germany, Andrews was at the Cosmos Club having lunch, when he ran into Charles Sheldon, in Naval Intelligence. Sheldon wanted Andrews in China for the war. Andrews went and while there crossed the Gobi Desert in a "motor car." He got an idea, and after returning to the United States in 1919, Andrews began organizing a series of explorations eventually conducted between 1922 and 1928, termed the Central Asiatic Expeditions. Their goal was to test Henry Fairfield Osborn's (and earlier, Ernst Haeckel's) "Out of Asia" theory of human origins (which contradicted Darwin's "Out of Africa" theory). Andrews's expeditions never gathered enough ancestral human remains to address Osborn's theory, but they did collect an astonishing number of fossils, including, most famously, dinosaur eggs. The Central Asiatic Expeditions made Andrews a household name. This fame propelled him to the directorship of the American Museum of Natural History (a position he, by disposition, was unsuited to hold). It has also caused many people to champion Andrews as the inspiration for Indiana Jones,[33] but Andrews has some stiff regional competition.

Gordon MacCreagh was born on August 8, 1886, in Perth, Indiana.[34] His parents were Scottish, and the family came to the United States so that his father, a natural historian and anthropologist, could study Native Americans. MacCreagh attended grade school in Perth, then was sent to live with his grandfather in Scotland. After graduation, he studied at Heidelberg University, where he got into a sword fight with a German student, whom he wounded. Believing he killed him and was going to be arrested, MacCreagh fled to India.

MacCreagh first worked on a barge in Calcutta, then on a tea plantation in Darjeeling. There, he began collecting Himalayan butterflies and other insects for a museum biologist. This experience begat others, which led to him collecting larger, more dangerous animals. MacCreagh hunted leopards and tigers, and once, to reach a collecting site, avoided lousy roads by driving his car from Bombay to Calcutta on the railroad track. He collected in the Malay Islands and Borneo, where he specialized in orangutans and big snakes. In Africa, he collected more animals than the circus would buy and went broke feeding them.

MacCreagh returned to India as an agent for British Intelligence. He also began to write. His first effort was a play that, after some success in India, was brought to New York, where it opened at the Amsterdam Theater. The production was soon shut down for featuring what the city's finest felt was excessive nudity. Stranded in New York, MacCreagh earned money playing his bagpipes and writing pulp fiction. In 1913 he sold his first short story, "The Brass Idol," to *Adventure* magazine.

MacCreagh also served in World War I, and two years later helped organize the Mulford Expedition, led by Dr. Henry H. Rusby, dean of Pharmacy at Columbia University. Their goal was to descend the Amazon River to collect plants to test for new pharmaceutical drugs. The expedition was, by almost any criterion, a disaster.[35] It was poorly planned, and soon after they began, the "eminent" scientists started bickering, then began leaving. MacCreagh stuck it out. His most interesting experience may have been drinking caapi—boiled extract from the vine *Banisteriopsis caapi* ("vine of the soul") and the plant *Psychotria viridis*, which contains the psychedelic compound N,N-dimethyltryptamine—and dancing the ritual Caapi Dance, intended to frighten away devils. The concoction took a day to wear off, and MacCreagh danced the whole time. He returned to the United States in 1922 and wrote *White Waters and Black*,[36] a no doubt true tale of his experiences on this expedition. Back in New York, he met Helen Komlosy (identified as "The Intrepid Exploress" in *The Last of Free Africa*), herself a traveler and expert shot. They married in June 1923.

In 1927 the MacCreaghs were in Abyssinia (Ethiopia), where they searched for the lost Ark of the Covenant. *Adventure* magazine funded the trip and published articles that MacCreagh posted during the expedition. The MacCreaghs met the Emperor Haile Selassie, who anointed MacCreagh a Knight of the Golden Star of Ethiopia. MacCreagh recounted his experiences in his bestselling book, *The Last of Free Africa*.[37] A few examples of MacCreagh's Groucho Marks–like quips serve to illustrate his popular style:

> It has been said that there are twenty-seven trees in Djibouti. I didn't see all of them.[38]

> Every one of our Indian wars originated in the bad blood growing out of the question of whether the settler has more right to the land he has obtained by purchase or by government grant, or whether the Indian who was born there has the better right.[39]

> . . . one cannot doubt the undisguised candor of men who each know that the other knows.[40]

This Land Is Your Land

No one will argue against the statement that it was field biologists who stocked our great museums, and therefore collected the data that formed our understanding of the species inhabiting—the biodiversity of—our continent. Nor will anyone argue with the notion that it was the field biologists who, with their knowledge of species, species distributions, and species interactions, laid—and continue to lay—the foundations for the fields of ecology, wildlife biology, conservation biology, and restoration ecology. This, in and of itself, is a complete legacy; it is the legacy—the only legacy—that outsiders and many insiders attribute to field biology.

But there is another legacy, just as tangible, that can be traced to American field biologists. They, and they alone, are primarily responsible for identifying and describing the 24 million km^2 of protected areas around the world set aside for natural processes, plus the environmental legislation that protects these areas and the plants and animals that inhabit them. Considering these lands in toto, Jim Furnish wrote (in verbiage reflecting the proud federal government employee he was):

> With courage and wisdom exhibited for more than a century, America has ambitiously set aside a vast complex of public lands—fully one-third of our country—for national parks, forests, wildlife refuges, and other public purposes. This penchant for conservation contrasted with much of the world's restive exploitation of natural resources. Now many other nations have belatedly embraced this concept. We see parks and reservations blossoming all over the globe. These lands have become increasingly important in the struggle to address an exploding population and mitigate the ensuing exploitation of resources. Management of our national forests mirrors, in microcosm, a much larger issue: how best to balance the needs of humans while adequately protecting the inherent integrity of the environment?[41]

Our National Park system, often called "America's best idea,"[42] was founded on the notion—embraced by our eighteenth-century field biologists—that this country has more to offer its citizens than opportunities to make money. The original criterion for parks and other set-asides was scenery. But then commercial logging threatened, and national forests—an idea borrowed but not copied from Europe—became protected. Later, habitat loss, unregulated commercial hunting, and weather vagaries caused game to become scarce, and wildlife refuges arose to protect this resource. Finally, as large unimpacted areas of nature became valued, wilderness areas became established. At about the same time, academic ecologists who had abandoned the "organisms have relationships" approach in favor of "relationships have organisms" were forced by the practicality of legislation designed to protect the environment, such as the Endangered Species Act, to once again consider organisms. (Hutchinson's view that relationships have organisms is scientifically sound but politically invisible. While the Endangered Species Act targets ecosystem *components*, we have no environmental laws specifically protecting specific dynamic ecological *processes* such as energy flow and nutrient recycling).

From this perspective, America's best idea was just the first in a series of best ideas that in total produced not just the first national parks, but also the first wildlife refuges, the first wilderness areas, and the first environmental legislation regulating their operations. As Furnish notes, once Americans invented these ideas, they became exports. From the time the Organic Act establishing the U.S. National Park Service passed in 1916 until 2011, the total number of nationally and internationally protected sites grew from nearly none to about 160,000, and the total area of protected terrestrial and marine sites grew from not much to over 24 million km^2 (fig. 49).[43] There are more protected natural areas on Earth now than ever before.[44] While the United States broadly divides its protected areas into parks, forests, wildlife refuges, and wilderness areas, more generally a protected area is simply "a clearly defined geographical space, recognized, dedicated and managed, through legal or other effective means, to achieve the long-term conservation of nature with associated ecosystem services and cultural values."[45]

In every place in every country where natural systems have been, or are being evaluated, some version of the old North American field biologists' game of "1) what's out there; 2) how does it fit together; 3) how do we save it; and 4) because we couldn't save it how can we put it back together again" gets played.[46] As Paul Dayton has noted, because an understanding of natural history is necessary at every stage of this process, none of this preservation happens without field biologists.

In the old days, field biologists went first, collecting and shipping specimens. Later they became more sophisticated as they catalogued and drew, and maybe even photographed their findings. Today, potential preserves are likely identified indoors, at a desk, using a computer with satellite digital imagery software such as Google Earth. After promising areas are identified, the field biologists get sent in, often organized into small teams, for example, under the Rapid Assessment Program developed by Conservation International, the Field Museum, and their partners. These results have been published by the University of Chicago Press under the Rapid Biological and Social Inventories book series, which by my count is up to twenty-seven publications.[47] Al Gentry, Ted Parker, and their colleagues were conducting a rapid assessment when they died in a plane crash in Ecuador in 1993.

If you accept, first, the argument that national parks (augmented by national forests, wildlife refuges, and wilderness areas) represent "America's best ideas"—ideas so good that they have been exported globally; and accept,

Total number of designated protected areas

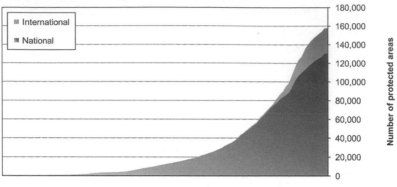

Total area of designated protected areas

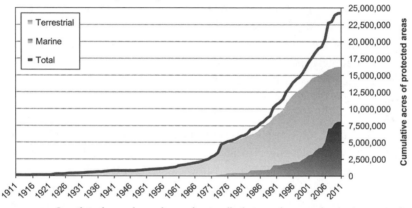

FIGURE 49. Growth in the number and area of nationally designated protected areas since 1916, when the U.S. National Park Service was established.

second, the idea that the massive environmental legislation passed in the United States in the 1960s and 1970s, which continues today (ideas so good that they, too, could be considered the "best"), complemented or created these set-asides; and realize that such set-asides and legislation could not have happened, and could not continue to happen, without the critical information provided by field biologists; then you cannot help but conclude that one legacy of American field biology is the global system of preserves and environmental protection legislation that we have come to embrace, and perhaps take for granted. This view brings a whole new perspective to, and a whole new appreciation for, the work of field biologists.

Follow this line of reasoning to its conclusion, and you make a startling discovery. America's best ideas render an America the opposite of itself. That is, to the world, America is viewed as the epitome of Adam Smith's notion of capitalism. But isn't the logical extension of national parks, national wildlife refuges, federal wilderness areas, and federal environmental legislation, as well as parallel efforts by states—America's best ideas—anti-capitalism? Not anti-capitalism in the sense of Soviet-style socialism or alternative economic/ social systems, but in the sentiment of the old-time field naturalists—that this country is about so much more to its citizens than money; about so much more than making a buck. We treat our families this way, why not treat our country the same?

In the movie *A Beautiful Mind*, based on the life of Noble Prize–winner John Nash, there is a scene where Nash, played by Russell Crowe, and four male friends are in a bar, when a statuesque blonde and her four female friends enter. Nash's friends each plot to "get" the blonde, and one of them evokes Adam Smith, "the father of modern economics. 'In competition individual ambition serves the common good.' " Nash considers and then replies, "Adam Smith needs revision. If we all go for the blonde we block each other. Not a single one of us is going to get her. So then we go for her friends, but they will all give us the cold shoulder because nobody likes to be second choice. But what if no one goes for the blonde? We don't get in each other's way, and we don't insult the other girls. . . . Adam Smith said, 'The best result comes from everyone in the group doing what's best for themselves.' Right? That's what he said. Incomplete. Incomplete. Because the best result will come from everyone in the group doing what's best for himself, and the group. . . . Governing dynamics, gentlemen, governing dynamics. Adam Smith was wrong."

Of course, there has been, and continues to be, resistance to this "good for the group" idea from über-capitalists (best represented historically by U.S. Representative "Uncle" Joe Cannon's line in the sand: "Not one [more] cent for scenery!"[48]). Wallace Stegner realized the conflict between the forces of self and the forces of cooperation and community.[49] He acknowledged that the Adam Smith–like rugged individual "really fired our imaginations, and still does . . . our principal folk hero, in all his shapes . . . is essentially anti-social." The lone cowboy or mountain man we can identify with; the "cooperating, stolid, dependable farmer—the settler or clod buster—is booed or scorned." (The Clint Eastwood movie *The Outlaw Josey Wales*[50] is much about making this transition from sodbuster to folk hero.) In reality, Stegner knew life is hard and unpredictable, and to survive, people need the support—physically and emotionally—of family and friends (where Josey Wales's story eventually ends).[51] "The whole history of mankind is social, not individual."[52] And in what may have been a time of despair, Stegner wrote:

> We are the unfinished product of a long becoming. In our ignorance and hunger and rapacity, in our dream of a better material life, we laid waste the

continent and diminished ourselves before any substantial number of us began to feel, little and late, an affinity with it, a dependence on it, an obligation
toward it as the indispensable source of everything we hope. . . . Land gave
Americans their freedom. It also gave them their egalitarianism, their democracy, their optimism, their free-enterprise capitalism, their greed, and their
carelessness. It is an ambiguous and troubling legacy.[53]

While we expect resistance to the idea of land set-asides from capitalists,
there has also been resistance from environmentalists. One of Stegner's former students, Terry Tempest Williams,[54] has deeply criticized wilderness areas
by challenging the "simple stories of innocent preservationists saving pristine
places."[55] She has instead, similar to George Catlin in the eighteenth century
and rightly so, focused on the fate of Native Americans who occupied these
lands, and the conflicts that arose among whites in setting aside these areas.
It seems, as with predator control and clear-cuts, there will always be controversy surrounding environmental protection—for example, the Pacific Northwest Forest Plan is once again being reconsidered.[56] We do not question these
concerns but can ask, on the flip side, where would we be without these wilderness areas? Given that we cannot answer the simple question "How much
biodiversity is enough?,"[57] isn't too much protection better than not enough, at
least until we are able to answer this question?[58] And can't we partially correct
the mistakes of the past by bringing indigenous peoples back into the conservation conversation?[59] Further, there are other challenges to set-asides that
must be addressed. For example, should we now be managing plant and animal populations in wilderness areas, especially in the face of climate change?[60]
We require strong leadership and vision to address these issues. No one should
say the old-timers got it all correct; no one does, the first time.

The legacy of British naturalists, so well known, is the theory of evolution through natural selection, developed independently by Charles Darwin
and Alfred Russel Wallace, and bull-doggedly fought for by Thomas Henry
Huxley. The until-now unrecognized legacy of American field biologists,
composed largely of midwestern talent, is the expanding global network of
natural areas and environmental protection legislation. (And who would
have guessed the important roles that schools such as Beloit College [the Yale
of the West!], Oberlin, and Iowa State played.)

You can debate whether the midwesterners highlighted here represent the
vanguard of this movement or are merely representative (that is, you can ask
the question did they drive these actions, or do they instead make a fine syn-
ecdoche?). I would argue the former, but I am likely biased. It's time to take a
closer look. Any movement that lasts a century and a half and involves scores

of people is easily underappreciated by an intellectual community comprising specialists focused on maximizing their number of published peer-reviewed journal articles. As some measure of this, compare how many people high-lighted here have biographies with how many deserve them. As any field bi-ologist will tell you (using exactly this terminology): sit your ass in front of a computer and your world becomes restricted to the world of the known and its derivatives. Extrapolate this thought to the world of historical biography, and you understand why the only people deemed worthy of biography are those who already have biographies. To correct this, field biologists will then tell you to do what they do—go out into the field and start looking and think-ing for yourself—and soon you realize the world becomes damn near limit-less. So, where are those biographies of the Muries, or Nelson, Allee, Hyman, Gabrielson, Gleason, Weaver, Grinnell, McAtee, Lacey, Dilg, and Wetmore? They must be written. These biologists were all there when it mattered. When the brains were perceived to be out East and the wilderness was out West, these people were alternately in both places; equally at home in a saddle or a desk chair. They saw, and when they thought about it and wrote about it, they were eloquent. Their history is largely the history of field biology in the United States, an endeavor whose commodity is so valuable it continues to be exported globally. With apologies to Wallace Stegner, it is, and remains, an unambiguous and hopeful legacy.

Notes

Preamble

1. As John Wesley Powell (2013, 6) wrote: "Their bronzed, hardy, brave faces come to me as they appeared in the vigor of life; their lithe but powerful forms seem to move around me; and the memory of the men and their heroic deeds, the men and their generous acts, overwhelms me with a joy that seems almost a grief, for it starts a fountain of tears."

2. Royte (2008).

3. E. Abbey, https://www.goodreads.com/author/quotes/37218.Edward_Abbey (accessed September 5, 2016).

4. From an Edward Abbey quote: "A landscape, like a man or a woman, acquires character through time and endurance" (Abbey 1989, 106).

5. Campbell (1988, 5).

6. Steinbeck and Ricketts ([1941] 1971, 28–29).

7. The style of this sentence is intentional and meant as homage to Norman Maclean.

8. Ted Parker and Al Gentry were beloved. For example (there are many), see Sullivan (1993); Conniff (2011a); and Remsen and Schulenberg (1997).

9. The memories of Ted Parker and Al Gentry survive in many different ways, but one of the most thoughtful is the Parker/Gentry Award for Conservation Biology sponsored by the Field Museum of Natural History. It is an annual award, funded by an anonymous donor, given to "an outstanding individual, team or organization whose efforts are distinctive and courageous and have had a significant impact on preserving the world's natural heritage, and whose actions and approaches can serve as a model to others" (see parkergentry.fieldmuseum.org). In 2001 I was fortunate to receive this award.

10. Mittermeier (2011, 9–13).

11. Gallagher (2013).

12. Weinberg (2000).

13. Bailey (2008, 33).

14. Murie (1957).

15. Schaller (1980, 258).

16. Iltis (1996); for a full consideration of Al Gentry, see eulogies by T. M. Barkley, J. S. Miller, W. H. Lewis, E. Forero, M. Plotkin, O. Phillips, P. H. Raven, and R. Rueda in the same volume, pp. 433–60, available at https://www.jstor.org/stable/pdf/2399988.pdf.

17. Powell (2013, 5).

18. Steinbeck and Ricketts ([1941] 1971, 6–7).

19. Schaller (1980).

20. Townsend ([1839], 1999, xix).

21. Schaller (1980, 77).

22. Cherrie (1930, 44–46).

23. Stein (2001), photographs opposite p. 174.

24. Pauley (2000, 51): "By the early 1880s more naturalists were working in and from Washington than any other place in the country. . . . A few of these younger Washington naturalists, most notably the Yale-educated geologist Clarence King, moved in these circles. But many more—including Hayden, Powell, and sociologist/botanist Lester F. Ward—had come from nondescript Midwestern backgrounds. Baird, in his search for assistants willing to work on the frontiers, had accepted a number of rough, independent men. These men had, to a significant extent, educated themselves before arriving in Washington. Creative but in certain respects naïve, they continued to rely on reading and discussion for their intellectual development."

25. "Fish Basics," Florida Museum, https://www.flmnh.ufl.edu/fish/education/questions /questions.html (accessed January 4, 2016).

26. Reptile Database, http://www.reptile-database.org/db-info/SpeciesStat.html (accessed January 4, 2016).

27. Bovbjerg (1988); for a full history, see Lannoo (2012).

28. Missouri Botanical Garden, http://www.mobot.org/mobot/archives/fulltext_images.asp (accessed August 22, 2014).

29. Kohler (2002).

30. It is surprisingly difficult to track down the source of this quote. See, for example, http:// www.goodreads.com/quotes.

31. Vitt (2013).

32. Dayton (2008).

33. There is some doubt about this. See Anonymous (2014); and Barrows, Murphy-Mariscal, and Hernandez (2016).

34. Greene (1997, 2013); Crump (2000, 2005, 2015); Sagarin and Pauchard (2012)—we are all going to miss Rafe.

35. Wilson (1994); Graham, Parker, and Dayton (2011); see also Sagarin and Pauchard (2012).

36. Box and Draper (1987, 424).

37. Stegner (1999, 137); for an alternative view, one that I do not disagree with, see https:// www.hcn.org/no-national-parks-are-not-americas-best-idea (accessed October 17, 2017).

38. Woody always wrote from a man on a box car's perspective as an observer going through the world; see http://www.youtube.com/watch?v=yZx7xCK6yfo.

Introduction

1. Steinbeck and Ricketts ([1941] 1971, 1).

2. Born in Clarinda, Iowa, on December 23, 1902.

3. Maclean (1976, 127).

4. Dayton (2008).

5. For a beautifully succinct summary of the environmental movement in the United States, see Kline (1997).

6. Lorimer (2015) proposes that because human influence is now everywhere on the planet, the term "wildlife" should be used instead of "nature." He offers an interesting concept, one that Stephen Forbes would have appreciated. I am probably too weathered to buy fully into this notion—I have a hard time visualizing prairie plants as wildlife. But, good for Lorimer. Ideas that ask us to reconsider what we do have value whether or not we choose to accept them.

7. For excellent descriptions of the explorers and explorations prior to Lewis and Clark's Voyage of Discovery, see Bakeless (1950) and Savage (1979).

8. For example, Stark (2014, 102) dismisses Nuttall because he used "his musket barrel as a spade to dig up the rootballs of plants, hopelessly clogging the weapon's muzzle with mud and rendering it useless in times of danger." First, consider Nuttall's contributions to our understanding on botany, his primary field. Then consider that in 1832 he wrote the definitive bird identification guide of its day (my copy of *A Popular Handbook of the Birds of the United States and Canada* is the 1911 edition published by Little, Brown, and Company, in Boston, Massachusetts). Few have contributed as much to our understanding of the world in the context of their times as Nuttall. Second, as with Ferdinand Hayden a half-century later, everyone, including the Native Americans, knew that if Nuttall wasn't a fool, he was so single-minded in his pursuit of specimens that he was harmless. Nuttall didn't require a musket for his own protection, nor did his companions likely consider counting on him—even with an operational firearm—for theirs.

9. There are a number of remarkable books written about natural historians and their museums, including Goetzmann (1986, 2000), Windsor (1991), Conn (1998), Welch (1998), Pauley (2000), Wallace (2000), Asma (2001), Blom (2002), Kohler (2002, 2006), and Conniff (2011b).

10. But see Wulf (2015); Rebok (2014).

11. Egerton (2012, 121–25).

12. Humboldt (1856).

13. Brought to my attention by a friend, U.S. Forest Service cinematographer Steven Dunsky. In Washburn (1978, 18–19), we read that the name "Cosmos Club" was proposed by Garrick Mallory and that its competitors were the "Saturn Club" and the "Kosmos Club." However, the inspiration for the name is never given, and I believe Steve is on to something.

14. These surveys have generated a large literature, including Bartlett (1962), Goetzmann (1991), Tidball (2004), and Moore (2006).

15. Preston (2007).

16. Egerton (2012) and Kingsland (2005).

17. Court (2012, 103).

18. Egerton (2015).

19. Kirksey (2015) eschews the historical baseline for an ahistorical one, and asks how organisms find emergent opportunities that either enhance or diminish their community assemblages. He posits, rather than intervene with the goal of achieving historical baselines (which may no longer be achievable), why not just watch these changes and learn from them?

The Foundation

1. Evans (1997, 3).

2. Bergeron (1989, xxiv).

3. Bakeless (1950, chaps. IV–VI).

4. Golay and Bowman (2003, 58).

5. Golay and Bowman (2003, 58–59).

6. Golay and Bowman (2003, 99).

7. Golay and Bowman (2003, 185–86).

8. Golay and Bowman (2003, 102).

9. Golay and Bowman (2003, 65).

10. Golay and Bowman (2003, 120–22).

11. Golay and Bowman (2003, 225).

12. Sterling (1997b).

13. Hornaday (1886, 373).

14. Hornaday (1886, 374, 375).

15. Bakeless (1950, chap. VII).

16. Cook (1997).

17. Cook (1997). The complete passage is as follows and can be found in Michelant (1865): "Neantmoins nous descendismes ce iour en quatre liex pour les arbres qui y estoyent tres-beaux, et de grande odeur, et trouuasmes que c'estoyent Cedres, Yfs, Pins, Ormeaux, Blancs, Fresnes, Saulx, et plusteurs autres à nous incogneus, tous neant moins sans fruit. Les terres oùbiln'y a point do bois sont tresbelles et toutes plaines de poids, de raisin blanc at rouge ayant la fleur blanche dessus, des fraizes, meures, froment sauuage comme seigle qui semble y auoir esté semé et labouré, et ceste terre est de meilleure temperature qu'aucune qui se puisse voir et de grande chaleur, l'on y voit oiseaux, en somme il n'y a faute d'autre chose que de bons ports."

18. Greenberg (2014, 46–47).

19. Bakeless (1950, 104).

20. Golay and Bowman (2003, 146–47).

21. Golay and Bowman (2003, 248).

22. Golay and Bowman (2003, 287–88).

23. Bakeless (1950, chap. XI).

24. Golay and Bowman (2003, 128).

25. Golay and Bowman (2003, 130).

26. Golay and Bowman (2003, 251–52).

27. As noted by Hayden ([1872] 2006), in the Great Plains, the source of a river is frequently larger than its mouth.

28. Golay and Bowman (2003, 238).

29. Golay and Bowman (2003, 377–78).

30. Sterling (1997a).

31. Bergeron (1989, xxvi).

32. Stegner ([1954] 1992, 374).

33. Johnsgard (1989).

34. Pick and Sloan (2004, 40–41).

35. Lottinville, introduction, in Nuttall ([1819] 1980, x).

36. Evans (1997, 11–12).

37. Golay and Bowman (2003, 315–16).

38. Brinkley (2009, 297).

39. Benson (1988, ii).

40. Savage (1979, chap. 7).

41. Evans (1997, 5), unnamed source.

42. Golay and Bowman (2003, 358–59).

43. DeVoto (1947).

44. Graustein (1967, 68–70); Savage (1979, 152, 176).

45. Bradbury ([1817] 1986).

46. Evans (1997, 13).

47. DeVoto (1947, 22).

48. DeVoto (1947, 23).

49. DeVoto (1947, 23).

50. Golay and Bowman (2003, 275–76).

51. Stegner ([1954] 1992, 374).

52. In a letter dated September 15, 1844, *The Expeditions of John Charles Frémont*, entry 86, pp. 365–68, https://archive.org/stream/expeditionsofjoh01fr/expeditionsofjoh01fr_djvu.txt (accessed September 8, 2016).

53. Nicollet (1993, 24–25).

54. Beidleman (2006, 209).

55. Graustein (1967, 72).

56. Savage (1979, 90n6).

57. Benson (1988); Evans (1997).

58. Benson (1988, iii).

59. Evans (1997, viii).

60. Evans (1997, viii).

61. Evans (1997, 14).

62. Evans (1997, 13).

63. Evans (1997, 16).

64. Benson (1988, i).

65. Benson (1988, ix).

66. Francisco Vázquez de Coronado, letter to the king of Spain, October 20, 1541. See https://www.tshaonline.org/handbook/online/articles/ryl02 (accessed October 13, 2014).

67. Cooper ([1827] 1987, 11).

68. Cather ([1918] 2008, 13).

69. Examining tree rings, David W. Stahle and Edward R. Cook inferred climate conditions during the 1805 Pike Expedition and the 1818 Long Expedition, and suggest how extraordinarily dry the Great Plains were during these times. See http://sites.nationalacademies.org/cs/groups/pgasite/documents/webpage/pga_086004.pdf (accessed February 26, 2017).

70. Benson (1988, xiv).

71. Benson (1988, xv).

72. Weaver (1954, 3).

73. Irving ([1832] 2013, 34).

74. Irving (1832, 30).

75. From 1815 to 1817, Prince Maximilian led a pioneering expedition to southeast Brazil. His book *Reise nach Brasilien* introduced images of Brazilian Indians to Europe. It was eventually translated into several European languages and is recognized as a great early contribution to the knowledge of Brazil. The original (in German) can be downloaded at https://archive.org/details/reisenachbrasili02wied.

76. DeVoto (1947, 17).

77. Witte and Gallagher (2008, 3–4).

78. Fenn (2014, 184–85) offers a fantastic look at the Mandan people. About the Welsh Indians, she writes: "The Madoc myth . . . had been promulgated by a bevy of Elizabethan courtiers,

cartographers, scholars, and theorists to challenge Spain's claims to America. . . . The hunt for the Welsh Indians had moved westward with Anglo-American settlement. By 1794, its focus had narrowed to two Indian peoples: the Apaches of the Southwest and the Mandans of the upper Missouri." Catlin (2004, 183) describes the Mandans: "Their singular and peculiar customs [including their dependence on agriculture] have raised an irresistible belief in my mind that they have had a different origin, or are of a different compound of character from any other tribe that I have yet seen, or that can be probably seen in North America. . . . [They appear to be] an amalgam of civilized and savage; and that the absence of all proof of any recent proximity of a civilized stock that could in any way have been engrafted upon them. . . . [These facts] lead the mind back in search of some more remote and rationale cause for such striking singularities; and in this dilemma, I have been almost disposed to enquire whether here may not be found, yet existing, the remains of the *Welsh colony*—the followers of Madoc."

79. Nicollet (1993, 4).

80. Nicollet (1993, 6).

81. Nicollet (1993, 9).

82. Savage (1979, 241).

83. Nicollet (1993, 15).

84. Maclean (1993, 164).

85. The buffalo referred to here is the American bison (*Bison bison*), subspecifically, the Plains bison (*B. b. bison*).

86. Nicollet (1993, 6).

87. McHugh (1972, 248).

88. Pagnamenta (2012, 44).

89. DeVoto (1947, 21), suggests this idea then rejects it in an endnote.

90. DeVoto (1947, 228).

91. Turner (1935, 9).

92. The Ordinance of 1787, created by an act of Congress, established the "Northwest Territory"—land that today includes the states of Michigan, Ohio, Indiana, Illinois, and Wisconsin, as well as a portion of northeastern Minnesota. This legislation set the stage for the form of the present-day United States by establishing that the country would expand by admitting new states to the union rather than by extending the western borders of existing states. Further, it prohibited slavery in this territory, establishing the Ohio River as the boundary between free and slavery regions from the Appalachian Mountains to the Mississippi River, setting the stage, three-quarters of a century later, for the Civil War (Turner 1935, 79–89).

93. Turner (1935, 109).

94. Turner (1935, 10).

95. Turner (1935, 81).

96. Turner (1935, 81–82).

97. Turner (1935, 82).

98. Turner (1935, 109).

99. Turner (1935, 14).

100. Turner (1935, 17–18).

101. Turner (1935, 185).

102. Turner (1935, 11).

103. Turner (1935, 86).

104. Turner (1935, 251–52).

105. Turner (1935, 113).

106. Overfield (1993, 4–5).

107. Turner (1935, 152).

108. Turner (1935, 152–53).

109. According to Oelschlaeger (1991, 134): "Emerson's *Nature* is not a philosophical inquiry. Instead it is a literary exercise designed to lay a pre-established belief in God on rational, rather than scriptural, footing. . . . For Emerson a wilderness experience allows a person's mind to first discover a reflection of itself, then use this knowledge to confirm God's existence."

110. Emerson (1836, iii).

111. My information is based on Gura (2007).

112. Gura (2007, 142).

113. Gura (2007, 269). While most people link Thoreau with Emerson under the Transcendentalist umbrella, Oelschlaeger (1991, 150) notes an important difference: "There can be no easy equation between Thoreauvian and Emersonian views of natural entities. For Thoreau they exist in and for themselves, whereas for Emerson they are ultimate commodities, provided by a benevolent God for his most perfect creation."

114. Gura (2007, 266).

115. Oelschlaeger (1991, 135): Emerson felt "nature is mere putty in human hands, bestowed by God upon his most favored creation, man." Egerton (2012, xii) traces the modern tradition of natural theology—finding proof of God's wisdom in plants and animals—back to John Ray in the 1600s.

116. Emerson (1836, 3).

117. Shepard ([1927] 1961, 119).

118. Shepard ([1927] 1961, 117–18).

119. Shepard ([1927] 1961, 148).

120. Shepard ([1927] 1961, 107–8).

121. Emerson (1836, 109).

122. Shepard ([1927] 1961, 77).

123. Oelschlaeger (1991, 139).

124. To identify species within these types, scientists use the Latinized binomial system of species names developed by Linnaeus (e.g., *Canis familiaris* for the domestic dog, *Canis latrans* for the coyote, *Canis lupus* for the gray wolf, etc.), and a standardized hierarchical classification scheme (e.g., phylum Chordata, class Mammalia, order Carnivora, family Canidae, genus *Canis*, species *familiaris*), which reflects the evolutionary relationships of species, recognized after Darwin's 1859 publication, *On the Origin of Species*.

125. Wilson (2004, 3).

126. Should anyone doubt this individual-group split, I offer the following from Thoreau-admirer Sigurd Olson's essay "The Red Squirrel," which appears in Olson (1984, 155): "The relationship of such vital factors as temperature and humidity to the phenomenon of hibernation has interested me not at all. Gestation periods, stomach analysis, and population cycles are closed books to me. Not once have I been tempted to contribute an article on them to the scientific journals, for what I have learned about squirrels and their ways is of no importance to anyone but myself."

127. Shepard ([1927] 1961, 219).

128. Here I rely on Dupree (1959).

129. Rodgers (1944, 78).

130. Savage (1979, 187).

131. DeVoto (1947, 183); derived from Irving (1836).

132. Graustein (1967, 132–52).

133. Torrey's published account of James's collections marks the first time an American botanist classified plants based on the new "natural system" (Ewan 1950, 14).

134. Dupree (1959, 74).

135. Dupree (1959, 76).

136. Dupree (1959, 83).

137. Pick and Sloan (2004, 44–45); see also Egerton (2012, 126).

138. Rodgers (1965, 89).

139. Nisbet (2009).

140. Gifford (1972).

141. Dupree (1959, 156).

142. Dupree (1959, 158).

143. DeVoto (1942, 471) said about Frémont on the occasion of his nomination to be the first presidential nominee of the newly formed Republican Party: Frémont "so jeopardized the Republican Party. . . . He was worse than a fool, he was an opportunist, an adventurer, a blunderer on a truly dangerous scale. . . . [H]e made a play for every opportunity that would serve John Charles Frémont, regardless of its effect on the United States." I believe the Republicans have now managed to Trump Frémont.

144. See Philbrick (2003).

145. Rodgers (1965, 159).

146. Rodgers (1965, 256).

147. Dupree (1959, 159).

148. Rodgers (1965, 249).

149. Rodgers (1965, 235–36).

150. Rodgers (1965, 147).

151. Dupree (1959, 212).

152. Dupree (1959, 207).

153. Dupree (1959, 214–15).

154. Dupree (1959, 325).

155. Rodgers (1949, 79–80).

156. Rodgers (1949, 81).

157. Rodgers (1949, 860).

158. Rodgers (1949, 119).

159. Rodgers (1949, 349).

160. Bailey (2008, xiii).

161. Rodgers (1949, 92).

162. Bailey (2008, 13).

163. Bailey (2008, 7).

164. Rodgers (1949, 198–99).

165. Meine (1988, 214, 296); Nash (2001, 194); Bailey (2008, 1).

166. Minteer (2006, 46, 49, 50); Bailey (2008, 1).

167. Rivinus and Youssef (1992, 57).

168. Rivinus and Youssef (1992, 58–59).

169. Rivinus and Youssef (1992, 44).

170. Rivinus and Youssef (1992, 27, 28).

171. Sterling (1977, 2).

172. Baird (1857, 1858, 1859).

173. Rivinus and Youssef (1992, 83).

174. Rivinus and Youssef (1992, 60).

175. Rivinus and Youssef (1992, 120–21).

176. Rivinus and Youssef (1992, 145).

177. Sterling (1977, 3).

178. Dupree (1959, 150–54).

179. Dupree (1959, 199).

180. Missouri Botanical Garden, http://www.missouribotanicalgarden.org/about/additional
-information/our-mission-history.aspx (accessed September 11, 2016).

181. Oberlin College was the first, admitting women in 1837. "Oberlin History," https://new
.oberlin.edu/about/history.dot (accessed September 12, 2016).

182. Yale Peabody Museum of Natural History, http://peabody.yale.edu/about-us/mission
-history (accessed September 12, 2016).

183. Greenberg (2014, 185).

184. Greenberg (2014, 186).

185. Sterling (1977, 6).

186. Sterling (1977, 290).

187. Osgood (1943).

188. Brinkley (2009, 107).

189. Foster (1994).

190. Rodgers (1944, 13).

191. See fig. 5 on p. 000, right-hand side of photograph, for a similar racial arrangement during the 1877 Hooker-Gray Expedition.

192. Stegner ([1971] 2014, 274). Stegner's description appears to be backed by a photograph of
the King camp reproduced in Goetzmann (2000, 632). It is also supported by the photograph of
Asa Gray and Joseph Hooker visiting the USGS Survey team at La Veta Pass, Colorado in 1877
(fig. 5; p. 35).

193. There are many published histories of the Hayden, Powell, King, and Wheeler surveys.
One of the most succinct can be found in Goetzmann (2000, 390–601).

194. Sterling (1977, 12).

195. Brinkley (2009, 107).

196. Sterling (1977, 24).

197. Sterling (1977, 32).

198. Sterling (1977, 42, 44).

199. Sterling (1977, 56).

200. Sterling (1977, 55).

201. Sterling (1977, 56, 63).

202. Brinkley (2009, 214).

203. Sterling (1977, 79); see also Schmidly, Tydeman, and Gardner (2016).

204. Vernon Bailey has only recently got the attention he deserves (Kohler, 2008).

205. Sterling (1977, 66).

206. Sterling (1977, 73).

207. Field Museum, https://www.fieldmuseum.org/about/history (accessed September 12, 2016).

208. Stoddard (1969, 140).

209. For a full consideration of Chicago in the context of American expansion and settlement, see Cronon (1991).

210. Sterling (1977, 95).

211. Sterling (1977, 101–5).

212. Sterling (1977, 91).

213. Sterling (1977, viii).

214. Miller (1928, 214–15).

215. Miller (1928, 406); Sterling (1977, 67).

216. Miller (1928, 407–8); Brinkley (2009, 214).

217. Sterling (1977, 166).

218. Sterling (1977, 242).

219. Sterling (1977, 173).

220. Sterling (1977, 178).

221. Sterling (1977, 277).

222. Sterling (1977, 71).

223. Brinkley (2009, 298–99).

224. Sterling (1977, 266).

225. Sterling (1977, 257).

226. Sterling (1977, 81).

227. Sterling (1977, 268).

228. Sterling (1977, 267).

229. Sterling (1977, 309).

230. Glover (1992, 134): "It was [Mardy] . . . who convinced [Olaus] in 1945 to resign from the Biological Survey to become halftime director of the Wilderness Society." Glover calls Murie a superstar (139).

231. Sterling (1977, 241).

232. Sterling (1977, 280).

233. Sterling (1977, 269).

234. Sterling (1977, 262).

235. Sterling (1977, 269–70).

236. Sterling (1977, viii).

237. Sterling (1977, 307).

238. Ambrose (2004, xiii–xiv).

239. Brinkley (2009, 163–64).

240. Brinkley (2009, 99).

241. Roosevelt ([1920], 2009, 19).

242. Brinkley (2009, 131).

243. Brinkley (2009, 130).

244. Brinkley (2009, 132–33). Many writers would be better served if they had a field biologist proofread manuscript drafts. Brinkley implies western Iowa as the site of the 1932 heath hen extinction (which in fact occurred on Martha's Vineyard, in Massachusetts) and of the decline of Attwater's prairie chicken (which are found on the Gulf Coastal Plain of Texas, an hour west of Houston).

245. Brinkley (2009, 134).

246. Brinkley (2009, 137).

247. Brinkley (2009, 145).

248. Brinkley (2009, 143–46).

249. Brinkley (2009, 146).

250. Brinkley (2009, 151).

251. Brinkley (2009, 153).

252. Brinkley (2009, 170–71).

253. Brinkley (2009, 169, 162–63).

254. Brinkley (2009, 174, 180–81).

255. In 2004, two of these books, *Hunting Trips of a Ranchman* and *The Wilderness Hunter*, were published with an introduction by Stephen Ambrose by the Modern Library, New York.

256. Brinkley (2009, 186).

257. Brinkley (2009, 196–97).

258. Brinkley (2009, 197–98).

259. Brinkley (2009, 198–200).

260. Cherrie (1930, 255).

261. Roosevelt ([1920], 2009, 15).

The Natural Historians

1. Pauley (2000, 51): "By the early 1880s more naturalists were working in and from Washington than any other place in the country. . . . [M]any . . . including Hayden, Powell, and sociologist/botanist Lester F. Ward—had come from nondescript Midwestern backgrounds. Baird, in his search for assistants willing to work on the frontiers, had accepted a number of [these] rough, independent men."

2. Cassidy (2000, 34–35).

3. Cassidy (2000, 35).

4. Cassidy (2000, 35–38).

5. Cassidy (2000, 42).

6. Cassidy (2000, 45).

7. Cassidy (2000, 56).

8. Cassidy (2000, 61–65).

9. Cassidy (2000, 68–69).

10. Cassidy (2000, 77).

11. Cassidy (2000, 78).

12. Cassidy (2000, 97–98).

13. Cassidy (2000, 98).

14. Cassidy (2000, 114).

15. Delo (1998, 295); No sentence I'd read while researching for this book jarred me like this one. Who thinks of nineteenth-century cavalry playing baseball? It's true, as Bowman (2000/2001) details. In the 1850s, baseball emerged and was a favorite pastime among Union soldiers during the Civil War (36). Postwar soldiers brought baseball with them to their frontier outposts (37). Games were played between companies, and between groups of soldiers and visitors. Custer's Seventh Cavalry and the Tenth Cavalry, a black unit, embraced baseball (42). While detailed accounts of many of these frontier contests exist, few record outcomes between black and white units. I do not wonder why.

16. Hayden ([1872] 2006, 74).

17. Hayden ([1872] 2006, 111–12).

18. Hayden ([1872] 2006, 122).

19. Hayden ([1872] 2006, 129).

20. Hayden ([1872] 2006, 155).

21. Hayden ([1872] 2006, 124).

22. Hayden ([1872] 2006, 455).

23. Hayden ([1872] 2006, 171).

24. Hayden ([1872] 2006, 456).

25. Hayden ([1872] 2006, 257).

26. Stegner ([1954] 1992, 11).

27. Stegner ([1954] 1992, 15).

28. Stegner ([1954] 1992, 16).

29. Stegner ([1954] 1992, 16–17).

30. As Stegner ([1954] 1992, 17) indicates: "Losing one's right arm is a misfortune; to some it would be a disaster, to others an excuse. It affected Wes Powell's life about as much as a stone fallen into a swift stream affects the course of the river. With a velocity like his, he simply foamed over it."

31. Stegner ([1954] 1992, 17).

32. Stegner ([1954] 1992, 6).

33. Powell (2013, 6).

34. Stegner ([1954] 1992, 6).

35. Stegner ([1954] 1992, 123).

36. Stegner ([1954] 1992, 240–41).

37. Stegner ([1954] 1992, 248–49).

38. Benson (1996, 125).

39. Benson (1996, 240).

40. This account has been extracted from Schlachtmeyer (2010).

41. Schlachtmeyer (2010, 112).

42. Schlachtmeyer (2010, 24).

43. Schlachtmeyer (2010, 31–33).

44. Schlachtmeyer (2010, 121).

45. This account is taken largely from Lewis (2012).

46. Lewis (2012, 7–8).

47. Lewis (2012, 11–34).

48. Lewis (2012, 37).

49. Lewis (2012, 135).

50. Lewis (2012, 55, 85).

51. Lewis (2012, 203–25).

52. Sterling (1977, 13).

53. Lewis (2012, 160–68).

54. Many books have been written about the Harriman Expedition and its luminaries, including Goetzmann and Sloan (1982), Burroughs, Muir et al. ([1901] 1986), Grinnell (1995), and Grinnell (2007).

55. Lewis (2012, 228–29).

56. Lewis (2012, 255–56).

57. Lewis (2012, 69).

58. Most of the information in the following two paragraphs comes from the Smithsonian's website: http://vertebrates.si.edu/fishes/ichthyology_history/ichs_colls/goode_g_brown.html (accessed October 17, 2017).

59. Two biographies of Hornaday appeared in consecutive years: Bechtel (2012) and Dehler (2013).

60. Dehler (2013, 18).

61. Dehler (2013, 19).

62. Bechtel (2012, 82).

63. Dehler (2013, 20).

64. Dehler (2013, 20).

65. Dehler (2013, 19).

66. Dehler (2013, 22).

67. Dehler (2013, 22).

68. Despite all of the campus construction since, as of mid-May 2016, when I last saw it, the plaque has remained in its original spot.

69. Dehler (2013, 22).

70. Dehler (2013, 23–25).

71. Dehler (2013, 26–27).

72. Dehler (2013, 28–29).

73. Dehler (2013, 31–36).

74. Dehler (2013, 37–40).

75. Dehler (2013, 42).

76. Hornaday (1929).

77. Dehler (2013, 52); see also Shell (2004).

78. Dehler (2013, 58–59).

79. Bechtel (2012, 5).

80. Bechtel (2012, 24); Dehler (2013, 61).

81. Bechtel (2012, 24).

82. Dehler (2013, 61–62).

83. Bechtel (2012, 43–44).

84. Bechtel (2012, 45).

85. Dehler (2013, 63).

86. Bechtel (2012, 39).

87. Dehler (2013, 64).

88. Because Hornaday killed these animals to demonstrate to the public their conservation status, Shell (2004, 103) calls the bison group, "a killing that denied itself."

89. Bechtel (2012, 49–50).

90. Dehler (2013, 64).

91. Norman Maclean indicates that Wag Dodge, foreman of the 1949 Smokejumper crew in Mann Gulch, operated this way. So does Maya Lin (2000, 9), who wrote: "I think with my hands."

92. Dehler (2013, 67).

93. Dehler (2013, 68).

94. Dehler (2013, 69).

95. Dehler (2013, 70); Bechtel (2012, 146).

96. Bechtel (2012, 147).

97. The good people of the city of Buffalo are not unaware of the name Bison. At the time of Hornaday, their professional baseball team, begun in 1877, was named the Bisons, as is their current Triple A team, a Toronto Blue Jays' farm club.

98. Bechtel (2012, 147).

99. Dehler (2013, 71–73).

100. Dehler (2013, 74–75).

101. Bechtel (2012, 155).

102. Bechtel (2012, 156).

103. Bechtel (2012, 172).

104. Cherrie (1930, 3).

105. Cherrie (1930, 134).

106. Cherrie (1930, 4–5).

107. Cherrie (1930, 7).

108. Cherrie (1930, 157).

109. Cherrie (1930, 129).

110. Cherrie (1930, 73–74).

111. Cherrie (1930, 83–84).

112. Cherrie (1930, 92–94).

113. Cherrie (1930, 158).

114. Cherrie (1930, 214).

115. Cherrie (1930, 101–2).

116. Cherrie (1930, 113–25).

117. Cherrie (1930, 141); see also Millard (2006).

118. Cherrie (1930, 248–49).

119. Although Kermit Roosevelt did not attribute these words, they come from the poem "He Has Gone," by Gilbert (1911, 27).

120. Cherrie (1930, vi).

121. Much of the information here was adapted from "Alexander Wetmore," https://www.pwrc.usgs.gov/resshow/perry/bios/wetmorealexander.htm (accessed October 17, 2017) and Ripley and Steed (1987).

122. Ripley and Steed (1987, 598).

123. And Wetmore would come to write Ridgway's obituary in the *Biographical Memoirs of the National Academy of Sciences* 15:57–101.

124. Ripley and Steed (1987, 598–99).

125. Wetmore (1915, 1918, 1919).

126. Ripley and Steed (1987, 599).

127. Volume I published in 1965; Volumes II and III published in 1972.

128. Ripley and Steed (1987, 607).

129. Wetmore (1926).

130. Ripley and Steed (1987, 609).

131. Ripley and Steed (1987, 600).

132. Ripley and Steed (1987, 611).

133. Ripley and Steed (1987, 611).

The Ecologists

1. Egerton (2012, 199).

2. From this perspective, it is no surprise to discover a book about "conservation after nature" written by a Brit (Lorimer 2015).

3. Golly (1993, 2).

4. Egerton (2015) combines the Chicago plant and animal schools. I chose to split them because the topics they covered and the trajectories of their descendants are so distinctly different. For example, Chicago's Shelford collaborated not with the descendants of Chicago's Cowles, but rather with Nebraska's Clements to write *Bio-Ecology*.

5. Much of this chapter is derived from Croker (2001).

6. Croker (2001, 14).

7. Croker (2001, 14).

8. Croker (2001, 17–43).

9. Croker (2001, 52).

10. Stegner ([1954] 1992, 22) calls Vasey a "bona fide and able naturalist . . . whose name still persists on the maps in the little curtain of maidenhair and redbud and ivy called Vasey's Paradise, deep in Marble Canyon on the Colorado."

11. Croker (2001, 54).

12. Croker (2001, 58).

13. Croker (2001, 65).

14. Croker (2001, 61–62).

15. Croker (2001, 66–67).

16. Croker (2001, 68).

17. Croker (2001, 69).

18. Croker (2001, 71).

19. Croker (2001, 71–72).

20. Croker (2001, 72).

21. Croker (2001, 78).

22. Croker (2001, 77).

23. Croker (2001, 76).

24. Croker (2001, 155).

25. Croker (2001, 73).

26. Croker (2001, 129).

27. Croker (2001, 83).

28. Croker (2001, 84).

29. Croker (2001, 90).

30. Croker (2001, 92).

31. Leopold Center for Sustainable Agriculture, http://www.leopold.iastate.edu/strips-research -team (accessed September 13, 2016).

32. Croker (2001, 101).

33. Croker (2001, 102).

34. Croker (2001, 111).

35. Croker (2001, 124).

36. Croker (2001, 125).

37. Croker (2001, 128).

38. Croker (2001, 134).

39. Croker (2001, 137).

40. Croker (2001, 138).

41. Croker (2001, 140).

42. Croker (2001, 145).

43. Croker (2001, 154).

44. Croker (2001, 144).

45. Croker (2001, 158).

46. Croker (2001, 156).

47. Croker (2001, 165).

48. Croker (2001, 167–68).

49. I extracted much of the information here from Tobey (1981) and Overfield (1993).

50. Tobey (1981, 10).

51. Dupree (1959, 394).

52. Tobey (1981, 10–11).

53. Tobey (1981, 11).

54. Tobey (1981, 10–11).

55. Tobey (1981, 14–15).

56. Tobey (1981, 17).

57. Kingsland (2005, 144).

58. Tobey (1981, 37).

59. Cittadino (1997).

60. Voigt (1980).

61. Voigt (1980, 318–19).

62. Voigt (1980, 319).

63. Voigt (1980, 320).

64. Voigt (1980, 321).

65. Voigt (1980, 320).

66. Weaver and Clements (1938).

67. With a nod toward Norman Maclean.

68. Clements (1905).

69. Cassidy (2007, 55).

70. Cassidy (2007, 53).

71. Rodgers (1944, 6).

72. Rodgers (1944, 10).

73. Rodgers (1944, 11).

74. Rodgers (1944, 13).

75. Foster (1994, 227–28).

76. Sterling (1977, 11).

77. Rodgers (1944, 24–25).

78. Rodgers (1944, 26).

79. Rodgers (1944, 29).

80. Rodgers (1944, 33).

81. Rodgers (1944, 47).

82. Rodgers (1944, 49).

83. Rodgers (1944, 50).

84. Rodgers (1944, 61).

85. Rodgers (1944, 64).

86. Rodgers (1944, 83).

87. Rodgers (1944, 102).

88. Rodgers (1944, 110).

89. Rodgers (1944, 126).

90. Rodgers (1944, 151).

91. Rodgers (1944, 158).

92. Rodgers (1944, 159).

93. Rodgers (1944, 173).

94. Cassidy (2007, 25).

95. Cassidy (2007, 13).

96. Cassidy (2007, 15).

97. Cassidy (2007, 15).

98. Cassidy (2007, 17).

99. Cassidy (2007, 19).

100. Cassidy (2007, 20).

101. Cassidy (2007, 21).

102. Cassidy (2007, 22–23).

103. Cassidy (2007, 24).

104. Cassidy (2007, 26).

105. Cassidy (2007, 27).

106. Cassidy (2007, 28); Sears (1964).

107. Cassidy (2007, 28).

108. Cassidy (2007, 28–29).

109. Cassidy (2007, 29).

110. Cassidy (2007, 31).

111. *Hillsdale Daily News*, http://www.newspapers.com/newspage/32599574/ (accessed October 17, 2017).

112. Cassidy (2007, 34).

113. Cassidy (2007, 30).

114. Cassidy (2007, 57).

115. Cassidy (2007, 6).

116. Cassidy (2007, 47).

117. Cassidy (2007, 47).

118. Cassidy (2007, 48).

119. Cassidy (2007, 93).

120. Cassidy (2007, 55).

121. Cassidy (2007, 53).

122. Cassidy (2007, 56).

123. Cassidy (2007, 51).

124. Cassidy (2007, 45).

125. Cassidy (2007, 43).

126. Cassidy (2007, 43).

127. Cassidy (2007, 80–84).

128. Cassidy (2007, 81).

129. Cassidy (2007, 82).

130. Cassidy (2007, 83).

131. Cassidy (2007, 84).

132. Cassidy (2007, 90).

133. Cassidy (2007, 91).

134. Cassidy (2007, 94).

135. Cassidy (2007, 93).

136. Cassidy (2007, 8).

137. See McIntosh (1975), Nicholson (1990), and Nicholson and McIntosh (2002).

138. Dunkak (2007).

139. Dunkak (2007, 23).

140. Dunkak (2007, 21).

141. Kingsland (2005, 4).

142. McIntosh (1975).

143. Gleason (1920, 1922, 1925, 1929).

144. McIntosh (1975, 256–58).

145. Goodall (1952); also McIntosh (1975, 257).

146. Nicholson and McIntosh (2002); Ricklefs and Miller (2000, 523–24); http://www.mac millanlearning.com/catalog/static/whf/ricklefsmiller (accessed October 17, 2017).

147. Gleason (1917, 1926).

148. McIntosh (1975, 262).

149. Gleason (1939).

150. McIntosh (1975, 265).

151. Gleason and Cronquist (1963, 1964).

152. Gleason and Cronquist (1964, 82).

153. Gleason and Cronquist (1964, 92).

154. Gleason and Cronquist (1964, 197).

155. Gleason and Cronquist (1964, 211–12).

156. The majority of the information here derives from Parker (1938).

157. Parker (1939, 203).

158. Kohlstedt (1985, 6).

159. Parker (1939, 203, 204–5).

160. Parker (1939, 205).

161. Croker (1991, 4).

162. Croker (1991, 5).

163. Croker (1991, 7).

164. Croker (1991, 8).

165. Croker (1991, 10).

166. Croker (1991, 12).

167. Croker (1991, 12–13).

168. Croker (1991, 15).

169. Croker (1991, 16).

170. Croker (1991, 16).

171. Croker (1991, 18–19).

172. Croker (1991, 21).

173. Croker (1991, 24).

174. Croker (1991, 25–26).

175. Croker (1991, 28–29).

176. Croker (1991, 29–30).

177. Croker (1991, 33–34).

178. Croker (1991, 34).

179. Croker (1991, 35).

180. Croker (1991, 49).

181. Croker (1991, 50).

182. Croker (1991, 56).

183. Croker (1991, 71).

184. Croker (1991, 72).

185. Backes (1997, 85).

186. Clements and Shelford (1939).

187. Croker (1991, 90).

188. Croker (1991, 93).

189. Croker (1991, 94).

190. Croker (1991, 102–3).

191. Croker (1991, 103).

192. Croker (1991, 104).

193. Croker (1991, 101).

194. Croker (1991, 101).

195. Croker (1991, 110–11).

196. Croker (1991, 111).

197. Unless otherwise indicated, the material in this account comes from Schmidt (1957).

198. Schmidt (1957, 10).

199. Schmidt (1957, 11).

200. Schmidt (1957, 12).

201. Schmidt (1957, 13).

202. Schmidt (1957, 18).

203. Schmidt (1957, 23).

204. Schmidt (1957, 25–26).

205. Hyman and Hutchinson (1991).

206. Hyman and Hutchinson (1991, 105).

207. Hyman and Hutchinson (1991, 107).

208. Hyman and Hutchinson (1991, 111).

209. This biographic sketch is derived from my longer work (Lannoo 2010); I retain the original references; Hedgpeth (1978a, 1–2).

210. Rodger (2002, xv).

211. Hedgpeth (1978a, 5).

212. Rodger (2002), xxi.

213. Tamm (2004, 10).

214. Allee (1923a,b).

215. Hedgpeth (1978a, 5).

216. Carol Steinbeck was a remarkable person. Quick-witted, it was her idea to name her husband's most famous novel *The Grapes of Wrath*.

217. Tamm (2004, 13).

218. Hedgpeth (1978a, 25).

219. Hedgpeth (1978a, 24–25).

220. Although the existence of this letter has been called into question (Benson 2010), it has made the rounds among those who know. Jim Carlton kindly sent me a scan of this letter, dated December 2, 1931, which Joel Hedgpeth had previously sent to him.

221. Hedgpeth (1978a, 26–30).

222. Rodger (2002, 38–39).

223. Steinbeck (1939).

224. DeMott (1989, 106); Rodger (2002, xxxvii).

225. DeMott (1989, 106).

226. Ricketts (2006, 34).

227. Benson (1984, 445).

228. Rodger (2002, xl–xli).

229. Ricketts (2006, 46).

230. Rodger (2002, 118).

231. Steinbeck and Ricketts ([1941] 1971).

232. Steinbeck and Wallsten (1989, 223).

233. Rodger (2002, xlv).

234. Rodger (2002, xlvi).

235. Hedgpeth (1978b, 21).

236. Hedgpeth (1978a, 42); Bob de Roos was an award-winning journalist who later worked for many magazines over his illustrious career, including *Life*, *Sports Illustrated*, and *National Geographic*.

237. Rodger (2002, 210).

238. Hedgpeth (1978a, 42); from *San Francisco Chronicle*, "This World," 26 VII 70.

239. Frey (1963, 4).

240. Frey (1963, 3).

241. Russell (1940).

242. Frey (1963, 4).

243. Russell (1940, 11).

244. Sellery (1956).

245. Russell (1940, 14); Frey (1963, 4).

246. Russell (1940, 13); Sellery (1956, 18).

247. Russell (1940, 13).

248. Sellery (1956, 15).

249. Sellery (1956, 16).

250. Frey (1963, 4).

251. Sellery (1956, 39–41).

252. Frey (1963, 5).

253. Frey (1963, 5).

254. Juday (1915).

255. Welch (1944, 272).

256. Frey (1963, 5)

257. Edward A. Birge, in Symposium on Hydrobiology (1940, 46).

258. Frey (1963, 3).

259. Ralph M. Immell, in Symposium on Hydrobiology (1940, 23).

260. http://www.brainyquote.com/quotes/quotes/p/paulsamuel205550.html.

The Wildlife Biologists

1. Much of this account is taken from Goldman (1935).

2. Goldman (1935, 138).

3. Goldman (1935, 138).

4. Goldman (1935, 139).

5. Goldman (1935, 140).

6. Goldman (1935, 142).

7. Goldman (1935, 142).

8. Goldman (1935, 143–44).

9. Goldman (1935, 145).

10. Goldman (1935, 145).

11. Glover (1992) calls him grouchy.

12. Goldman (1935, 136).

13. Goldman (1935, 147).

14. Lendt (1989, 4).

15. Lendt (1989, 6).

16. Lendt (1989, 6); in Chase (1971), we learn that "Pat and Mike" stories are gathered under the title "The Fool Irishman Tales," which in England were termed "The Wise Men of Gotham" stories.

17. Lendt (1989, 6).

18. Lendt (1989, 15–16).

19. Lendt (1989, 7).

20. For example, Lendt (1989, 18), relates that late in life Darling recalled that Gifford Pinchot carried on a continuous battle for forest preservation against his superior, Secretary of the Interior Ballinger, whom Roosevelt had inherited from McKinley. This is incorrect; Roosevelt inherited Ethan A. Hitchcock. Richard A. Ballinger served Roosevelt as commissioner of the General Land Office from 1907 until 1908. In 1909 Ballinger served as Taft's secretary of the interior.

21. Lendt (1989, 10).

22. Yankton would later produce the Denver Bronco/Oakland Raider great Lyle Alzado, who, apparently, owed his professional football career to Ding Darling.

23. Lendt (1989, 11).

24. Lendt (1989, 13).

25. Lendt (1989, 12).

26. Lendt (1989, 14).

27. Lendt (1989, 18).

28. Lendt (1989, 18). This is likely another case of Darling's faulty memory. Errington was born in 1902, and in his autobiography, he says he did not learn to shoot until after he contracted polio in 1910 and was attempting to regain his strength. By 1910 Darling had been in Des Moines four years.

29. Lendt (1989, 59).

30. It was a long and confusing name for a newspaper and in 1916 *and Leader* was dropped.

31. Lendt (1989, 21).

32. Lendt (1989, 24).

33. Lendt (1989, 23).

34. Lendt (1989, 28).

35. Lendt (1989, 32).

36. Lendt (1989, 31).

37. Lendt (1989, 48).

38. Lendt (1989, 39).

39. Lendt (1989, 37).

40. Lendt (1989, 89).

41. Lendt (1989, 89).

42. Lendt (1989, 173).

43. Lendt (1989, 33).

44. Lendt (1989, 95).

45. Lendt (1989, 47).

46. Lendt (1989, 38).

47. Lendt (1989, 109).

48. Lendt (1989, 84).

49. Lendt (1989, 96).

50. Lendt (1989, 101).

51. Lendt (1989, 125).

52. Backes (1997, 64).

53. Lendt (1989, 49).

54. Lendt (1989, 63).

55. Lendt (1989, 61).

56. The term "welfare state" means different things to different people. For many Republican politicians, it denotes lazy people on the public dole. But as Otteson (2014, 1) describes, capitalism and socialism meet in the middle in "welfare-state-ism," which leaves property in private hands but regulated, with a system of progressive taxation and aid to poorer segments of society. I will add that such a system also provides common areas, such as parks, national forests and grasslands, wildlife refuges, and wilderness areas, for public enjoyment.

57. Lendt (1989, 63).

58. Lendt (1989, 64).

59. Lendt (1989, 65–67).

60. Lendt (1989, 69).

61. Lendt (1989, 70).

62. Lendt (1989, 65).

63. Kalmbach (1963).

64. Kalmbach (1963, 478–79).

65. Kalmbach (1963, 482).

66. Kalmbach (1963, 482).

67. Kalmbach (1963, 482).

68. For example, when considering the common names of my favorite group of North American frogs, Harper (1935), wrote: "The evidence at hand indicates that the range of *Rana capito* lies wholly within, and the range of *R. areolata* wholly without, that of [the gopher tortoise] *Gopherus polyphemus*. . . . Just as *capito* owes its common name to the Gopher Turtle with which it associates, so 'Crawfish Frog' is a fitting name for *areolata*, by reason of its appropriation of the crustacean's burrows for its own habitations. 'Northern Gopher Frog,' employed by some authors for *areolata* is scarcely a suitable name for a species that is not known to have any contact with the Gopher Turtle."

69. Kalmbach (1963, 484).

70. Kalmbach (1963, 485).

71. Lendt (1989, 70).

72. Herb Stoddard felt otherwise. In his biography (Stoddard 1969, 230), he seems genuinely

fond of McAtee and calls him "one of the most capable biological editors of his time." Paul Errington was also a trusted correspondent (Pritchard et al. 2006, 1412). These men were no fools; McAtee must have been desperate in either his quest for power or job security to resort to belittling Leopold.

73. Lendt (1989, 70–71).

74. Lendt (1989, 75–76).

75. Lendt (1989, 71).

76. Lendt (1989, 72–73).

77. Lendt (1989, 72).

78. Lendt (1989, 78).

79. Lendt (1989, 79). I own a sixteen-gauge Browning semiautomatic shotgun built in 1937 that has a shortened two-shot magazine designed in response to these regulations.

80. Lendt (1989, 81).

81. Lendt (1989, 80).

82. Lendt (1989, 103, 143).

83. Lendt (1989, 38).

84. Lendt (1989, 42).

85. Lendt (1989, 45).

86. Lendt (1989, 99–100).

87. Lendt (1989, 100).

88. Lendt (1989, 110).

89. Much of this account is from Reeves and Marshall (1985).

90. Lannoo (2012).

91. Lendt (1989, 72).

92. Reeves and Marshall (1985, 867).

93. Stoddard (1969, 3).

94. Stoddard (1969, 6).

95. Stoddard (1969, 61).

96. Stoddard (1969, 61–62).

97. Stoddard (1969, 69).

98. Stoddard (1969, 74–75).

99. Stoddard (1969, 83–84).

100. Stoddard (1969, 87).

101. Stoddard (1969, 87).

102. Stoddard (1969, 87–89).

103. Stoddard (1969, 108–10).

104. Stoddard (1969, 111–24).

105. Stoddard (1969, 122).

106. Stoddard (1969, 165).

107. Stoddard (1969, 176).

108. Stoddard (1969, 178–80, 183–84).

109. Stoddard (1969, 194).

110. Brinkley (2009, 134).

111. Little (2000).

112. Little (2000, 531).

113. Little (2000, 538).

114. Little (2000, 540–41).

115. For an excellent account of this portion of Murie's life, see Glover (1992). Murie never much liked Nelson's writing, "He is extremely critical and to the point, and so very correct in everything. I haven't enjoyed being with him much" (132). And later, "I wish though that he wouldn't jump on a fellow the way he does," Murie wrote. "He's got the whole outfit here in fear and trembling it seems. Believe me, he doesn't own my soul, only my labor, and I am not going to worry about pleasing him." (136).

116. Glover (1992, 134): Olaus intended to hire his brother Martin, but Martin succumbed to tuberculosis, so he hired Adolph.

117. Murie (1961, 3–4).

118. Murie (1957, 77, 78).

119. Murie (1957, 79); memorialized in John Denver's 1995 "Song for All Lovers" as follows: "I see them dancing, somewhere in the moonlight, somewhere in Alaska, somewhere in the sun."

120. Murie (1957, 96, 226).

121. Murie (1957, 80, 85).

122. Murie (1957, 80).

123. Murie (1957, 183).

124. Murie (1957, 182–83).

125. Murie (1957, 95).

126. Murie (1957, 109); Murie's shotgun was a double-barrel (Glover, 1992, 133), while Cherrie's was a tri-barrel.

127. Glover (1992, 133).

128. Murie (1957, 117, 134).

129. Murie (1957, 216).

130. Murie (1957, 134).

131. Murie (1957, 274).

132. Murie (1957, 211).

133. Murie (1957, 212).

134. Glover (1992, 135).

135. Glover (1989, 34).

136. Murie (1961, 7); also Murie (1944).

137. Murie (1961, 9–10).

138. Murie (1961, 10).

139. Goetzmann and Sloan (1982).

140. Murie (1961, 11).

141. Murie (1961, 13).

142. Murie (1944, 126–43).

143. Murie (1961, 149).

144. Errington (1973, 6).

145. Errington (1973, 6, 19).

146. Errington (1973, 6–7).

147. Errington (1973, 12).

148. Errington (1973, 13).

149. Errington (1973, 15).

150. Errington (1973, 17).

151. Errington (1973, 24).

152. Errington (1973, 27).

153. With a nod to John Prine.

154. Errington (1973, 37–39).

155. Errington (1973, 47).

156. Errington (1973, 34).

157. Errington (1973, 34).

158. Errington (1973, 35).

159. Errington (1973, 137).

160. Errington (1973, 137).

161. Errington (1973, 137–38).

162. Errington (1948, 341).

163. Nina told me this story, and subsequently I was able to relate it to Fred Errington, Paul's son. Fred said he'd always wondered how his father learned to dance, since he knew his parents' first date had been at a roadside bar with a band. I owe my friendship to both Nina Leopold Bradley and Fred Errington to the kindness of Curt Meine.

164. According to Carolyn, Errington did not just want a wife—he wanted an intellectual partner, someone who would be involved in his research. She was a skilled writer, herself, and edited both his scientific and popular manuscripts.

165. Leopold ([1933] 1986).

166. Lendt (1989, 59).

167. Lendt (1989, 60).

168. Sivils (2012, 18).

169. Pritchard et al. (2006).

170. This account derives from three sources. The first is a profile available online, published October 24, 2001, http://www.theoutdoorwire.com/features/225155 (hereafter Outdoor Wire 2011). The second is a short biography in conjunction with his art, http://www.mayberry fineart.com/artist/albert_hochbaum (accessed October 17, 2017). The third is his obituary, by Houston (1988).

171. Outdoor Wire (2011).

172. Outdoor Wire (2011, 2).

173. Outdoor Wire (2011).

174. Outdoor Wire (2011).

175. Houston (1988).

176. Outdoor Wire (2011).

177. Meine (1988, 416–17).

178. Meine (1988, 453).

179. Meine (1988, 458).

180. Houston (1988).

181. As with the Ricketts account, the information here derives from Lannoo (2010).

182. Meine (1988, 16).

183. Meine (1988, 18).

184. Meine (1988, 24).

185. Meine (1988, 35).

186. Meine (1988, 80–81).

187. Meine (1988, 93).

188. Meine (1988, 94).

189. Meine (1988, 121).

190. Meine (1988, 122–32).

191. Meine, in Tanner (1987, 40).

192. Meine (1988, 161).

193. Meine (1988, 198).

194. Meine (1988, 227–28).

195. Meine (1988, 225).

196. Meine (1988, 234).

197. Meine (1988, 247–48).

198. Meine (1988, 256).

199. Meine (1988, 256).

200. Meine (1988, 262).

201. Meine (1988, 266).

202. Meine (1988, 278).

203. Meine (1988, 278).

204. Meine (1988, 285).

205. Leopold ([1933] 1986, 423).

206. According to McCabe (1987, 11), Leopold impressed Dean Harry L. Russell and other members of the "Getaway Club," a group of men that met to present papers on outdoor, historical, and travel experiences.

207. Meine (1988, 307).

208. Lorbiecki (1996, 123).

209. Lorbiecki (1996, 185–86).

210. There is some confusion about whether the family first saw the Shack in 1935 or 1936. On May 14, 2008, I e-mailed Curt Meine: "In Tanner (1987, 170), Nina writes that the whole family saw the Shack for the first time in February 1936, but there is no 1936 Shack entry until March 1st. There is a February entry for 1935. Do you think Nina's 1936 is the wrong date (that it was 1935), or that Nina's February is the wrong month (that it was really March). For March 7th and 8th, 1936, 'Whole family' is written." Curt replied: ". . . my guess is that Nina (or the editors!) had the year mistaken, and the February 1935 date is the right one. There were differences in the memories of the children about their first visit(s). After interviewing the four surviving children separately, I had four divergent accounts! When I had them all together and explained, they threw their hands up and asked me to just use my best judgment from the evidence!"

211. Nina Leopold Bradley quoted in Tanner (1987, 170–71).

The Conservation Biologists

1. You can read about the latest conservation/preservation dustup (and it was a big one) in Nijhuis (2014).

2. Pinchot ([1947] 1988).

3. Pinchot ([1947] 1988, 1).

4. Pinchot ([1947] 1988, 82).

5. Pinchot ([1947] 1988, 10, 11).

6. Pinchot ([1947] 1988, 13).

7. Pinchot ([1947] 1988, 48).

8. Pinchot ([1947] 1988, 100).

9. Pinchot ([1947] 1988, 104).

10. Pinchot ([1947] 1988, 107–8).

11. Pinchot ([1947] 1988, 135).

12. Pinchot ([1947] 1988, 136).

13. Pinchot ([1947] 1988, 257).

14. Pinchot ([1947] 1988, 256).

15. Pinchot ([1947] 1988, 124).

16. Pinchot ([1947] 1988, 100).

17. Sutter (2002, 213).

18. Pinchot ([1947] 1988, 152–53).

19. Wilderness advocate Bob Marshall's father, Louis, was deeply influential in getting this school built (Glover 1986, 38).

20. Pinchot ([1947] 1988, 259).

21. Those wishing to know more about Hutcheson can go to the Francis Hutcheson Institute website: http://www.fhinst.co.uk/.

22. Muir ([1912/1913] 1965).

23. Muir ([1912/1913] 1965, 3).

24. Muir ([1912/1913] 1965, 29).

25. Muir ([1912/1913] 1965, 45).

26. Muir ([1912/1913] 1965, 59).

27. Muir ([1912/1913] 1965, 162).

28. Muir ([1912/1913] 1965, 186).

29. Muir ([1912/1913] 1965, 197).

30. Muir ([1912/1913] 1965, 216).

31. Teale ([1954] 2001, xi).

32. Teale ([1954] 2001, 175).

33. Muir (1916); this passage is from Teale ([1954] 2001, 75–76).

34. Teale ([1954] 2001, 99).

35. Teale ([1954] 2001, 102).

36. Teale ([1954] 2001, 106).

37. Teale ([1954] 2001, 162); Oelschlaeger (1991, 180), points out: "In contrast to Emerson, the mature Muir does not approach nature with an established belief in a transcendent God and then find in nature's beautiful panoply confirmation in that belief. Rather he actually finds divinity in wild nature. Second, Emerson is a theist, while for Muir, God is nature."

38. Teale ([1954] 2001, 235).

39. Teale ([1954] 2001, 246).

40. Teale ([1954] 2001, 266).

41. Teale ([1954] 2001, 272).

42. Teale ([1954] 2001, xvi).

43. Lendt (1989, 18).

44. From a session entitled "The Pinchot-Muir Split Revisited" held at the Society of American Foresters National Convention, in Portland, Oregon, on October 26, 2007, http://studylib.net /doc/8077109/the-pinchot-muir-split-revisited (accessed July 15, 2016).

45. Miller (2001, 141).

46. Egerton (2015, 12).

47. Sutter (2002, 73).

48. Sutter (2002, 73).

49. Sutter (2002, 74).

50. Sutter (2002, 74).

51. Sutter (2002, 74); see also Pearson (1922).

52. Dexter (1978).

53. Egerton (2015, 54–55).

54. 40 C.F.R. Chapter I, Subchapter C – Air Programs (Parts 50–97).

55. Pub. L. 88-577.

56. 40 C.F.R. Chapter I, Subchapters D, N, O (Parts 100–140, 401–471, 501–503).

57. 16 U.S.C. § 1531 et seq.

58. Again, see Nijhuis (2014).

59. Nijhuis (2014).

60. For example, see Meyer (1997).

61. Muir ([1912/1913] 1965).

62. Sutter (2002, 51).

63. Mitchell (1981, 48).

64. Pub. L. 59-209, 34 Stat. 225, 16 U.S.C. § 431–433.

65. As I write this, President Barack Obama has expanded Hawai'i's Papahānaumokuākea Marine National Monument to 1.5 million km², making it the largest protected area in the world. "New Protections for U.S. Lands and Waters," *Science* 353:971. As well, he designated the first Marine National Monument in the Atlantic Ocean, see Davis (2016).

66. Maclean (1993, 207).

67. 16 U.S.C. §471 et. seq.

68. "Short History of the Refuge System," U.S. Fish and Wildlife Service, http://www.fws.gov /refuges/history/over/over_hist-d_fs.html (accessed October 17, 2017).

69. Hornaday (1886).

70. 16 U.S.C., §703–708, 710–712.

71. "The Wilderness Act of 1964," Wilderness Connect, http://www.wilderness.net/nwps /legisact.

72. Cronon, in Sutter (2002, vii).

73. Backes (1997, 101).

74. Cronon, in Sutter (2002, x–xi).

75. Brinkley (2016).

76. Sutter (2002, 48).

77. Sutter (2002, 48–49).

78. Cronon, in Sutter (2002, xi).

79. Sutter (2002, 108–9).

80. Sutter (2002, 107).

81. Sutter (2002, 106).

82. Sutter (2002, 105).

83. Sutter (2002, 231).

84. Sutter (2002, 231).

85. "Founders," Wilderness Society, http://wilderness.org/bios/founders (accessed October 17, 2017).

86. Cronon, in Sutter (2002, xi).

87. Cronon, in Sutter (2002, vii).

88. Harvey (2005, 13).

89. Harvey (2005, 14–15).

90. Harvey (2005, 16).

91. Harvey (2005, 56, photo).

92. Harvey (2005, 17–18).

93. Harvey (2005, 20–21).

94. Harvey (2005, 30–31).

95. Harvey (2005, 47–48).

96. Harvey (2005, 50–51).

97. Harvey (2005, 123).

98. Harvey (2005, 123).

99. Harvey (2005, 123).

100. Harvey (2005, 125).

101. Sutter (2002).

102. Harvey (2005, 73–77).

103. Harvey (2005, 94–97).

104. Harvey (2005, 98).

105. Harvcy (2005, 100).

106. Harvey (2005, 100).

107. For an in-depth account of this controversy, see Harvey (1994).

108. Harvey (2005, 104).

109. Harvey (2005, 104).

110. Harvey (2005, 105).

111. Harvey (2005, 105).

112. Harvey (1994, 283).

113. Harvey (2005, 117).

114. Pinchot (1947, 124).

115. Harvey (2005, 117–18).

116. Harvey (2005, 82–83).

117. Harvey (2005, 83–86).

118. Harvey (2005, 126).

119. Harvey (2005, 129).

120. Harvey (2005, 128).

121. Harvey (2005, 126).

122. Harvey (2005, 137).

123. Harvey (2005, 172–73).

124. The best account of this story is found at https://www.nps.gov/parkhistory/onlinc_books /choh/admin_history/history4.htm, and outside of Rachel Carson's and Howard Zahniser's work might possibly be the most well-written article ever penned by a government employee (accessed October 17, 2017).

125. "Chesapeake and Ohio Canal: The Making of a Park," National Park Service, https:// www.nps.gov/parkhistory/online_books/choh/admin_history/history4.htm.

126. Harvey (2005, 175–76).

127. Harvey (2005, 186).

128. Harvey (2005, 187).

129. Harvey (2005, 188).

130. Harvey (2005, 219).

131. Harvey (2005, 237).

132. Harvey (2005, 244).

133. http://www.fws.gov/refuges/history/over/over_hist-d_fs.html.

134. Harvey (2005, 246).

135. Berry (2001, 93–94).

136. Sutter (2002, 230).

137. Harvey (2005, 82–83).

138. Backes (1997, 6–8).

139. Backes (1997, 10).

140. Backes (1997, 11).

141. Backes (1997, 21).

142. Backes (1997, 10, 11).

143. Backes (1997, 12).

144. Backes (1997, 13).

145. Backes (1997, 15).

146. Backes (1997, 22).

147. Backes (1997, 19).

148. Backes (1997, 20).

149. Backes (1997, 20).

150. Backes (1997, 23).

151. Backes (1997, 22–23).

152. Backes (1997, 33, 34).

153. Backes (1997, 36).

154. Backes (1997, 16).

155. Windsor (1991, 137–39).

156. Backes (1997, 37).

157. Backes (1997, 38).

158. Backes (1997, 37).

159. Backes (1997, 39).

160. Backes (1997, 41).

161. Backes (1997, 41).

162. Backes (1997, 43).

163. Backes (1997, 44).

164. Backes (1997, 45).

165. Backes (1997, 48).

166. Backes (1997, 48).

167. Backes (1997, 49, 50).

168. Backes (1997, 52).

169. Backes (1997, 57).

170. Backes (1997, 59).

171. Drummond (1905, 58).

172. Backes (1997, 57).

173. Backes (1997, 59).

174. Backes (1997, 67).

175. Backes (1997, 68).

176. Backes (1997, 59–60).

177. Backes (1997, 60).

178. Backes (1997, 66).

179. Backes (1997, 65).

180. Backes (1997, 63).

181. Backes (1997, 70).

182. Backes (1997, 78).

183. Backes (1997, 79).

184. Backes (1997, 79).

185. Backes (1997, 85).

186. Backes (1997, 89).

187. Backes (1997, 90).

188. Backes (1997, 91).

189. Backes (1997, 98).

190. Backes (1997, 96).

191. Backes (1997, 97).

192. Backes (1997, 99).

193. Backes (1997, 100).

194. Backes (1997, 101).

195. Backes (1997, 113).

196. Backes (1997, 114–15).

197. Backes (1997, 115).

198. Backes (1997, 122).

199. Backes (1997, 109).

200. Backes (1997, 124).

201. Backes (1997, 131).

202. Backes (1997, 131).

203. Backes (1997, 150–51).

204. Backes (1997, 162–63).

205. Backes (1997, 163).

206. Backes (1997, 170).

207. Backes (1997, 174).

208. Backes (1997, 179).

209. Backes (1997, 172, 182, 224).

210. Backes (1997, 192).

211. Backes (1997, 193).

212. Backes (1997, 188).

213. Backes (1997, 315).

214. Backes (1997, 213).

215. Backes (1997, 220).

216. Backes (1997, 285).

217. Backes (1997, 320–21).

218. Backes (1997, 296).

219. Backes (1997, 228).

220. Backes (1997, 233).

221. Backes (1997, 242).

222. Backes (1997, 254).
223. Backes (1997, 255).
224. Backes (1997, 280).
225. Backes (1997, 320, 330).
226. Backes (1997, 292).
227. Backes (1997, 311).
228. Backes (1997, 186).
229. Nash (1989, chap. 5).
230. Robinson (2005); Van Nuys (2015, 54).
231. Van Nuys (2015, 19).
232. Van Nuys (2015, 19).
233. Van Nuys (2015, 19).
234. Van Nuys (2015, 33).
235. Van Nuys (2015, 34).
236. Van Nuys (2015, 31).
237. Dunlap (1988, 38).
238. Van Nuys (2015, 43–44).
239. Van Nuys (2015, 67, 49–50).
240. Van Nuys (2015, 70).
241. Van Nuys (2015, 67).
242. Dunlap (1988, 49); Van Nuys (2015, 69).
243. Van Nuys (2015, 57).
244. Van Nuys (2015, 86).
245. Dunlap (1988, 65).
246. Van Nuys (2015, 85).
247. Dunlap (1988, 80).
248. Van Nuys (2015, 124).
249. Van Nuys (2015, 125).
250. Dunlap (1988, 49).
251. Van Nuys (2015, 91, 93).
252. Van Nuys (2015, 127).
253. Grinnell and Storer (1916).
254. Van Nuys (2015, 93).
255. Van Nuys (2015, 94).
256. Van Nuys (2015, 94).
257. Van Nuys (2015, 94–95).
258. Van Nuys (2015, 95).
259. Van Nuys (2015, 101).
260. Van Nuys (2015, 102–3).
261. Van Nuys (2015, 102).
262. Dunlap (1988, 59–60).
263. Dunlap (1988, 60).
264. Dunlap (1988, 60).
265. Van Nuys (2015, 108).
266. Van Nuys (2015, 109).
267. Van Nuys (2015, 109).

268. Van Nuys (2015, 111).

269. Van Nuys (2015, 111).

270. Van Nuys (2015, 111–12).

271. Van Nuys (2015, 112).

272. Gabrielson (1941, 208–10).

273. Van Nuys (2015, 114, 113).

274. Dan Grubner, personal communication.

275. Errington (2015, 208).

276. Leopold ([1933] 1986, 239).

277. Dunlap (1988, 71).

278. There is good evidence that Native Americans had burned prairies and savannas for millennia, prior to European settlement. Some have proposed that this led to strong selection for fire tolerance, especially among native plants. If true, it means that humans have long been exerting selection pressures on the North American biota. Viewed another way, in North America, non-human species have been evolving in response to human activities for, perhaps, over 10,000 years. For these species, the absence of fire becomes a greater challenge to survivorship than the continued presence of fire.

279. Dunlap (1988, 72–73).

280. Dunlap (1988, 73).

281. Dunlap (1988, 73).

282. Stoddard and Errington (1938).

283. Dunlap (1988, 73–74).

284. Dunlap (1988, 74–75).

285. Van Nuys (2015, 122).

286. Dunlap (1988, 74–75).

287. Murie (1940, 146).

288. Van Nuys (2015, 141–42).

289. Kalmbach (1963).

290. Dunlap (1988, 78–79).

291. Dunlap (1988, 79).

292. Dunlap (1988, 81).

293. Dunlap (1988, 115).

294. Van Nuys (2015, 132).

295. Van Nuys (2015, 133).

296. Van Nuys (2015, 122–23).

297. Van Nuys (2015, 121).

298. Van Nuys (2015, 141).

299. Van Nuys (2015, 150).

300. Van Nuys (2015, 150).

301. Van Nuys (2015, 155).

302. Van Nuys (2015, 156).

303. Van Nuys (2015, 161).

304. Souder (2000); Lannoo (2008).

305. Van Nuys (2015, 159).

306. Van Nuys (2015, 161).

307. Richard M. Nixon, Message to Congress, February 8, 1972; Van Nuys (2015, 166).

308. Van Nuys (2015, 167).

309. 7 U.S.C. §136–136y.

310. P.L. 92-516.

311. Van Nuys (2015, 183–84).

312. Van Nuys (2015, 184).

313. Van Nuys (2015, 191).

314. Van Nuys (2015, 210); with or without recovery plans, absent targeted extermination, some populations of wolves, cougars, and grizzlies are slowly recovering and reclaiming historic ranges. This is creating challenges for natural resources departments (*The Wildlife Professional* 8 [2014]: 14).

315. Van Nuys (2015, 283).

316. The debate continues today; see Goldfarb (2016).

317. Bacon (2012). The reader might also be interested in two professional reviews of Lyons and Graves (2014), by Holovich and Reyna (2016) and Wydeven (2016).

318. Furnish (2015, 188).

319. Sutter (2002, 56–57).

320. Sutter (2002, 60).

321. Sutter (2002, 62).

322. Sutter (2002, 63).

323. Sutter (2002, 63–64).

324. Sutter (2002, 64).

325. Sutter (2002, 64).

326. Sutter (2002, 68).

327. Sutter (2002, 69).

328. Sutter (2002, 69).

329. The impact of roads on wildlife is more nuanced than was imagined in the 1920s and 1930s; for example, see Ernst et al. (2016).

330. Sutter (2002, 70).

331. Sutter (2002, 85).

332. Sutter (2002, 87).

333. Sutter (2002, 252).

334. Sutter (2002, 253).

335. Sutter (2002, 253).

336. Furnish (2015, 44–45).

337. Furnish (2015, 188).

338. Furnish (2015, 188).

339. Furnish (2015, 65).

340. Sutter (2002, 201).

341. Sutter (2002, 201).

342. Furnish (2015, 188).

343. Furnish (2015, 188). For example, the U.S. Forest Service's actions in Alaska's Tongass National Forest serve as a particularly nasty example of resistance to change under the euphemism "continuity of purpose"; see Shoaf (2000); also Soderberg and DuRette (1988); and Mackovjack (2010, 271).

344. Furnish (2015, 191).

345. Furnish (2015, 2).

346. 42 U.S.C. 4321.

347. 140 C.F.R., Subchapters D, N, O (Parts 100–140, 401–471, 501–503).

348. 16 U.S.C. 35.

349. P.L. 94-588; Furnish (2015, 2).

350. Furnish (2015, 190–91).

351. Furnish (2015, 46).

352. Furnish (2015, 190–91).

353. Furnish (2015, 194).

354. Furnish (2015, 190).

355. Furnish (2015, 3).

356. Thomas (2015).

357. Furnish (2015, 11).

358. Furnish (2015, 151).

359. Furnish (2015, 151).

360. Furnish (2015, 160). In fact, it is still a political hotbed. Current clear-cutting in the Tongass National Forest is threatening the last remaining populations of the Alexander Archipelago wolf (*Canis lupus ligoni*). Why? Because Alaska is remote (out of sight, out of mind) and its federal politicians are powerful.

361. Furnish (2015, 161).

362. Furnish (2015, 3–4).

363. Furnish (2015, 189).

364. Furnish (2015, 197).

365. See, for example, Haber and Holleman (2013).

366. Nash (1989) and Martin and Franklin (2012).

The Restoration Biologists

1. Wildlands Network, http://www.wildlandsnetwork.org (accessed October 17, 2017).

2. Clewell and Aronson (2013, 3).

3. Galatowitsch (2012, 27–28).

4. Clewell and Aronson (2013, 163).

5. Court (2012, 75–77).

6. Court (2012, 122).

7. Court (2012, 97).

8. Court (2012, 97–98).

9. Court (2012, 99–100).

10. Sperry (1935).

11. Court (2012, 100).

12. Court (2012, 101).

13. Not in 1934, as Clewell and Aronson (2013, 163), claim.

14. Court (2012, 105).

15. Court (2012, 106).

16. Court (2012, 106), based on interviews by Court with Sperry, recorded between 1981 and 1992 (University of Wisconsin Oral History Collection, University of Wisconsin Archives, and the Arboretum Archival Files).

17. Court (2012, 106).

18. Court (2012, 93, 106–7).

19. Court (2012, 95).

20. Court (2012, 218–19); Cather ([1918] 2008, 13).

21. Court (2012, 110).

22. Court (2012, 111).

23. Court (2012, 137).

24. Which are still being used and lauded, see Logan et al. (2016).

25. Clewell and Aronson (2013, 4).

26. Stoddard (1969, 194).

27. Seager (2016).

28. Andrews ([1943] 2013, 2).

29. Andrews ([1943] 2013, 7).

30. Andrews ([1943] 2013, 36).

31. Andrews ([1943] 2013, 52).

32. Andrews ([1943] 2013, 98).

33. In particular, the Smithsonian Channel is fond of the Andrews theory, promoted in their episode *The Real Story: Indiana Jones* (http://www.smithsonianchannel.com/shows/the -real-story/679).

34. PulpFlakes, http://pulpflakes.blogspot.com/2012/08/gordon-maccreagh-adventurer-ex plorer.html (accessed October 17, 2017).

35. Although the expedition's scientists managed to collect a reasonable number of plants. See Brooklyn Botanic Garden, http://www.bbg.org/collections/herbarium_mulford (accessed October 17, 2017).

36. MacCreagh (1926).

37. MacCreagh (1928).

38. MacCreagh (1928, 20).

39. MacCreagh (1928, 40).

40. MacCreagh (1928, 83).

41. Furnish (2015, 7).

42. For an example, see Egan (2016); David Quammen, in his usual idiosyncratic way, calls Yellowstone "America's wild idea"; see National Geographic, http://www.nationalgeographic.com /magazine/2016/05/ (accessed October 17, 2017).

43. World Database on Protected Areas, available at https://www.iucn.org/theme/protected -areas/our-work/world-database-protected-areas (accessed October 17, 2017).

44. Julia Whitty, "There Are More Protected Places on Earth Now than Ever Before," *Mother Jones*, April 19, 2012, www.motherjones.com/blue-marble/2012/04/protected-planet (accessed October 17, 2017).

45. World Database on Protected Areas, available at https://www.iucn.org/theme/protected -areas/our-work/world-database-protected-areas.

46. It is curious that wildlife biologists do not now get the attention they once did. They should. Since the Federal Aid in Wildlife Restoration (Pittman-Robertson Act) of 1937 was passed, American sportsmen have contributed $56.9 billion dollars to conservation activities, including, last year alone, $823 million from Pittman-Robertson, $812 million from hunting license revenues, $686 million from fishing license revenues, and $624 million from the Dingle-Johnson Act (also called the Federal Aid in Sport Fish Restoration Act, which is also United States federal law 16 U.S.C. §§ 7770–777l). Globally, where funding such as this has gone to

habitat protection, species richness is more than 10 percent higher and within-species abundance is 14.5 percent higher, in comparison with non-protected areas.

47. University of Chicago Press, http://press.uchicago.edu/ucp/books/series/FM-RBSI.html (accessed October 17, 2017).

48. "Joseph Gurney Cannon Facts," http://biography.yourdictionary.com/joseph-gurney -cannon (accessed September 22, 2016).

49. Benson (1996, 22).

50. The Malpaso Company, Warner Brothers, Hollywood, California, 1976.

51. Benson (1996, 192).

52. Benson (1996, 319).

53. Benson (1996, 390).

54. Williams (2016).

55. Farmer (2016).

56. Cornwall (2016).

57. Oliver (2016).

58. Furthermore, set-asides don't always equate to protection, especially in the face of climate change; see Normile (2016).

59. See, for example, the special package of articles published in the Winter 2010 issue of *The Wildlife Professional*, published by the Wildlife Society. As well, see Brondizio and Le Tourneau (2016) and Hoagland (2016).

60. Kurth (2014).

References

Abbey, E. 1989. *A Voice Crying in the Wilderness (Vox Clamitas in Deserto): Notes from a Secret Journal*. New York: St. Martin's.

Allee, W. C. 1923a. "Studies in Marine Ecology: I. The Distribution of Common Littoral Invertebrates of the Woods Hole Region." *Biological Bulletin* 44:167–91.

———. 1923b. "Studies in Marine Ecology: III. Some Physical Factors Related to the Distribution of Littoral Invertebrates." *Biological Bulletin* 44:205–53.

Ambrose, S. E. 2004. Introduction to *Hunting Trips of a Ranchman and The Wilderness Hunter*, by Theodore Roosevelt. New York: Modern Library.

Andrews, R. C. (1943) 2013. *Under a Lucky Star*. Madison, WI: Borderland Books.

Anonymous. 2014. "Natural Decline: Few Biology Degrees Still Feature Natural History. Is the Naturalist a Species in Crisis?" *Nature* 508:7–8.

Asma, S. T. 2001. *Stuffed Animals and Picked Heads: The Culture and Evolution of Natural History Museums*. Oxford: Oxford University Press.

Backes, D. 1997. *A Wilderness Within: The Life of Sigurd F. Olson*. Minneapolis: University of Minnesota Press.

Bacon, T. 2012. "The Implementation of the Animal Damage Control Recovery Act: A Comment on Wildlife Service's Methods of Predatory Animal Control." *Journal of the National Association of Administrative Law Judiciary* 32. http://digitalcommons.pepperdine.edu/naalj/vol32/iss1/6.

Bailey, L. H. 2008. *Liberty Hyde Bailey: Essential Agrarian and Environmental Writings*. Edited by Z. M. Jack. Ithaca, NY: Cornell University Press.

Baird, S. F. 1857. Part 1. *General Report upon the Zoology of the Several Pacific Railroad Survey Routes*. Vol. VIII: *Mammals*. Washington, DC: Government Printing Office.

———. 1858. Part 2. *General Report upon the Zoology of the Several Pacific Railroad Routes*. Vol. IX: *Birds*. Washington, DC: Government Printing Office.

———. 1859. Part 3. *General Report upon the Zoology of the Several Pacific Railroad Routes*. Vol. X: *Reptiles*. Washington, DC: Government Printing Office.

Baird, S. F., and C. Girard. 1854. *A Catalogue of North American Serpents*. Washington, DC: Smithsonian Miscellaneous Collections, Number 2.

Bakeless, J. 1950. *America as Seen by Its First Explorers*. New York: Dover.

Barrows, C. W., M. L. Murphy-Mariscal, and R. R. Hernandez. 2016. "At a Crossroads: The Nature of Natural History in the Twenty-First Century." *BioScience* 66:592–99.

Bartlett, R. A. 1962. *Great Surveys of the American West.* Norman: University of Oklahoma Press.

Bechtel, S. 2012. *Mr. Hornaday's War: How a Peculiar Victorian Zookeeper Waged a Lonely Crusade for Wildlife That Changed the World.* Boston: Beacon Press.

Beidleman, R. G. 2006. *California's Frontier Naturalists.* Berkeley: University of California Press.

Benson, J. J. 1984. *John Steinbeck, Writer.* New York: Penguin.

———. 1996. *Wallace Stegner: His Life and Work.* New York: Penguin.

Benson, K. 2010. Review of *Leopold's Shack and Ricketts's Lab: The Emergence of Environmentalism,* by Michael J. Lannoo. *Journal of the History of Biology* 43:805–7.

Benson, M. 1988. *From Pittsburgh to the Rocky Mountains: Major Stephen Long's Expedition, 1819–1820.* Golden, CO: Fulcrum Press.

Bergeron, F. 1989. *The Journals of Lewis and Clark.* New York: Penguin.

Berry, W. 2001. *Arctic Refuge: A Circle of Testimony.* Minneapolis: Milkweed Press.

Blom, P. 2002. *To Have and to Hold: An Intimate History of Collectors and Collecting.* Woodstock, NY: Overlook Press.

Bovbjerg, R. V. 1988. "Status of the Iowa Lakeside Laboratory 1988: Recent History and Assessments." Unpublished manuscript [reviewed and amended by J. C. Downey, B. W. Menzel, and R. W. Cruden], 34 pp., plus appendices.

Bowman, L. 2000/2001. "Soldiers at Play: Baseball on the American Frontier." *Nine* 9:35–49.

Box, G. E. P., and R. Draper. 1987. *Empirical Model-Building and Response Surfaces.* New York: Wiley.

Bradbury, J. (1817) 1986. *Travels in the Interior of America in the Years 1809, 1810, and 1811.* Lincoln: University of Nebraska Press.

Brinkley, D. 2009. *The Wilderness Warrior: Theodore Roosevelt and the Crusade for America.* New York: HarperCollins.

———. 2016. *Rightful Heritage: Franklin D. Roosevelt and the Land of America.* New York: Harper.

Brondizio, E. S., and F-M. Le Tourneau. 2016. "Environmental Governance for All: Involving Local and Indigenous Populations Is Key to Effective Environmental Governance." *Science* 352:1272–73.

Burroughs, J., J. Muir et al. (1901) 1986. *Alaska: The Harriman Expedition, 1899.* New York: Dover.

Campbell, J., with B. Moyers. 1988. *The Power of Myth.* Edited by B. S. Flowers. New York: Doubleday.

Cassidy, J. G. 2000. *Ferdinand V. Hayden: Entrepreneur of Science.* Lincoln: University of Nebraska Press.

Cassidy, V. M. 2007. *Henry Chandler Cowles: Pioneer Ecologist.* Chicago: Kedzie Sigel Press.

Cather, W. S. (1918) 2008. *My Antonia.* Boston: Houghton Mifflin; repr., Read How You Want Classics Library.

Catlin, G. 2004. *North American Indians.* Edited by Peter Matthiessen. New York: Penguin.

Chase, R. 1971. *American Folk Tales and Songs.* New York: Dover Books on Music.

Cherrie, G. K. 1930. *Dark Trails: Adventures of a Naturalist.* New York: G. P. Putnam's Sons.

Cittadino, E. 1997. "Frederic Edward Clements." In *Biographical Dictionary of American and Canadian Naturalists and Environmentalists,* edited by K. B. Sterling, R. P. Harmond, G. A. Cevasco, and L. F. Hammond, pp. 156–58. Westport, CT: Greenwood Press.

Clements, F. E. 1905. *Research Methods in Ecology.* Lincoln, NE: University Publishing Company.

Clements, F. E., and V. E. Shelford. 1939. *Bio-Ecology*. London: John Wiley and Sons. My copy is an original, once housed in the library at Albion College, in Albion, Michigan.

Clewell, A. F., and J. Aronson. 2013. *Ecological Restoration: Principles, Values, and Structure of an Emerging Profession*. 2nd ed. Washington, DC: Island Press.

Conn, S. 1998. *Museums and American Intellectual Life, 1876–1926*. Chicago: University of Chicago Press.

Conniff, R. 2011a. "Dying for Discovery. Opinionator." *New York Times*, January 16.

———. 2011b. *The Species Seekers: Heroes, Fools, and the Mad Pursuit of Life on Earth*. New York: Norton.

Cook, P. L. 1997. "Cartier, Jacques." In *Biographical Dictionary of American and Canadian Naturalists and Environmentalists*, edited by K. B. Sterling, R. P. Harmond, G. A. Cevasco, and L. F. Hammond, pp. 140–42. Westport, CT: Greenwood Press.

Cooper, J. F. (1827) 1987. *The Prairie*. New York: Viking Penguin.

Cornwall, W. 2016. "Scientists Split on Oregon Old-Growth Forest Plan." *Science* 353:637.

Court, F. E. 2012. *Pioneers of Ecological Restoration: The People and Legacy of the University of Wisconsin Arboretum*. Madison: University of Wisconsin Press.

Croker, R. A. 1991. *Pioneer Ecologist: The Life and Work of Victor Ernest Shelford, 1877–1968*. Washington, DC: Smithsonian Institution Press.

———. 2001. *Stephen Forbes and the Rise of American Ecology*. Washington, DC: Smithsonian Institution Press.

Cronon, W. 1991. *Nature's Metropolis: Chicago and the Great West*. New York: Norton.

Crump, M. 2000. *In Search of the Golden Frog*. Chicago: University of Chicago Press.

———. 2005. *Headless Newts Make Great Lovers: And Other Unusual Natural Histories*. Chicago: University of Chicago Press.

———. 2015. *Eye of Newt and Toe of Frog, Adder's Fork and Lizard's Leg: The Lore and Mythology of Amphibians and Reptiles*. Chicago: University of Chicago Press.

Davis, Julia Hirschfeld. 2016. "Obama Creates Atlantic Ocean's First Marine Monument." *New York Times*, September 15, http://www.nytimes.com/2016/09/16/us/politics/obama-to-create-atlantic-oceans-first-marine-monument.html?_r=0.

Dayton, P. K. 2008. "Why Nature at the University of California?" *The NRS Transect* 26:7–14.

Dehler, G. J. 2013. *The Most Defiant Devil: William Temple Hornaday and His Controversial Crusade to Save American Wildlife*. Charlottesville: University of Virginia Press.

Delo, D. M. 1998. *The Yellowstone, Forever! The Fascinating Story of Our First National Park*. Helena, MT: Kingfisher Books.

DeMott, R. 1989. *Working Days: The Journals of The Grapes of Wrath*. New York: Penguin.

Denver, J. 1995. "A Song for All Lovers." Lyrics © 1995. BMG Rights Management U.S., LLC, Reservoir Media Management, Inc.

DeVoto, B. 1942. *The Year of Decision: 1846*. New York: St. Martin's Griffin.

———. 1947. *Across the Wide Missouri*. New York: Houghton Mifflin.

Dexter, R. W. 1978. "History of the Ecologists' Union—Spin-Off from the ESA and Prototype of The Nature Conservancy." *Ecological Society of America Bulletin* 59:146–47.

Drummond, W. H. 1905. *The Voyageur and Other Poems*. New York: G. P. Putnam's Sons.

Dunkak, H. M. 2007. *Knowledge, Truth, and Service: The New York Botanical Garden, 1891 to 1980*. Lanham, MD: University Press of America.

Dunlap, T. R. 1988. *Saving America's Wildlife: Ecology and the American Mind, 1850–1990*. Princeton, NJ: Princeton University Press.

Dupree, A. H. 1959. *Asa Gray: American Botanist, Friend of Darwin*. Baltimore: Johns Hopkins University Press.

Egan, T. 2016. "We're Winning!" *New York Times*, August 19.

Egerton, F. N. 2012. *Roots of Ecology: Antiquity to Haeckel*. Berkeley: University of California Press.

———. 2015. *A Centennial History of the Ecological Society of America*. New York: CRC Press.

Emerson, R. W. 1836. *Nature*. Boston: James Munroe and Company.

Ernst, R., M. Hölting, K. Rodney, V. Benn, R. Thomas-Caesar, and M. Wegmann. 2016. "A Frog's Eye View: Logging Road's Buffer Against Further Biodiversity Loss." *Frontiers in Ecology* 14:353–55.

Errington, P. L. 1948. "In Appreciation of Aldo Leopold." *Journal of Wildlife Management* 12:341–50.

———. 1973. *The Red Gods Call*. Ames: Iowa State University Press.

———. 2015. *Of Wilderness and Wolves*. Iowa City: University of Iowa Press.

Evans, H. E. 1997. *The Natural History of the Long Expedition to the Rocky Mountains, 1819–1820*. New York: Oxford University Press.

Ewan, J. A. 1950. *Rocky Mountain Naturalists*. Denver: University of Denver Press.

Farmer, J. 2016. "Meditations on Conservation: An Environmental Activist Urges a Renewal of the American National Park Idea." *Science* 352:1283.

Fenn, E. A. 2014. *Encounters at the Heart of the World: A History of the Mandan People*. New York: Hill and Wang.

Foster, M. 1994. *Strange Genius: The Life of Ferdinand Vandeveer Hayden*. Niwot, CO: Roberts Rinehart Publishers.

Frey, D. G. 1963. "Wisconsin: The Birge-Juday Era." In *Limnology in North America*, edited by D. G. Frey, pp. 3–54. Madison: University of Wisconsin Press.

Furnish, J. 2015. *Toward a Natural Forest: The Forest Service in Transition (A Memoir)*. Corvallis: Oregon State Press.

Gabrielson, I. N. 1941. *Wildlife Management*. New York: Macmillan.

Gallagher, T. 2013. *Imperial Dreams: Tracking the Imperial Woodpecker through the Sierra Madre*. New York: Atria Books.

Galatowitsch, S. M. 2012. *Ecological Restoration*. Sunderland, MA: Sinauer Associates.

Gifford, G. E., Jr. 1972. "Edward Fredrick Leitner (1812–1832), Physician-Botanist." *Bulletin of the History of Medicine* 46. Reprinted in *Broward Legacy* 27:5–23.

Gilbert, J. S. (the Isthmian Kipling). 1911. *Panama Patchwork: Poems by James Stanley Gilbert*. 2nd ed. Panama City, Panama: Panama Star and Herald Company.

Gleason, H. A. 1917. "The Structure and Development of the Plant Association." *Bulletin of the Torrey Botanical Club* 44:463–81.

———. 1920. "Some Application of the Quadrat Method." *Bulletin of the Torrey Botanical Club* 44:463–81.

———. 1922. "On the Relation between Species and Area." *Ecology* 3:158–62.

———. 1925. "Species and Area." *Ecology* 6:66–74.

———. 1926. "The Individualistic Concept of the Plant Association." *Bulletin of the Torrey Botanical Club* 53:1–20.

———. 1929. "The Significance of Raunkiaer's Law of Frequency." *Ecology* 8:299–326.

———. 1939. "The Individualistic Concept of the Plant Association." *American Midland Naturalist* 21:92–110.

Gleason, H. A., and A. Cronquist. 1963. *Manual of Vascular Plants of the Northeastern United States and Adjacent Canada*. New York: D. Van Nostrand Company.

———. 1964. *The Natural Geography of Plants*. New York: Columbia University Press.

Glover, J. M. 1986. *A Wilderness Original: The Life of Bob Marshall*. Seattle: Mountaineers Books.

———. 1989. "Thinking Like a Wolverine: The Ecological Evolution of Olaus Murie." *Environmental Review* 13:29–45.

———. 1992. "Sweet Days of a Naturalist: Olaus Murie in Alaska, 1920–1926." *Forest and Conservation History* 36:132–40.

Goetzmann, W. H. 1986. *New Lands, New Men: America and the Second Great Age of Discovery*. New York: Viking Press.

———. 1991. *Army Exploration in the American West, 1803–1863*. Austin: Texas State Historical Association.

———. 2000. *Exploration and Empire: The Explorer and the Scientist in the Winning of the American West*. Austin: Texas State Historical Association.

Goetzmann, W. H., and K. Sloan. 1982. *Looking Far North: The Harriman Expedition to Alaska, 1899*. Princeton, NJ: Princeton University Press.

Golay, M., and J. S. Bowman. 2003. *North American Exploration*. Hoboken, NJ: John Wiley and Sons.

Goldfarb, B. 2016. "No Proof That Predator Culls Save Livestock, Study Claims." *Science* 353: 1080–81.

Goldman, E. A. 1935. "Edward William Nelson—Naturalist, 1855–1934." *The Auk* 52:135–48.

Golly, F. B. 1993. *A History of the Ecosystem Concept in Ecology: More than the Sum of Its Parts*. New Haven, CT: Yale University Press.

Goodall, D. W. 1952. "Quantitative Aspects of Plant Distribution." *Biological Review* 27:194–245.

Graham, M. H., J. Parker, and P. K. Dayton. 2011. *The Essential Naturalist: Timeless Readings in Natural History*. Chicago: University of Chicago Press.

Graustein, J. E. 1967. *Thomas Nuttall, Naturalist: Explorations in America, 1808–1841*. Cambridge, MA: Harvard University Press.

Gray, A., and J. D. Hooker. 1880. "The Vegetation of the Rocky Mountain Region and a Comparison with That of Other Parts of the World." *U.S. Geological and Geographical Survey of the Territories*. Bulletin VVI:1 77.

Greenberg, J. 2014. *A Feathered River Across the Sky: The Passenger Pigeon's Flight to Extinction*. New York: Bloomsbury.

Greene, H. W. 1997. *Snakes: The Evolution of Mystery in Nature*. Berkeley: University of California Press.

———. 2013. *Tracks and Shadows: Field Biology as Art*. Berkeley: University of California Press.

Grinnell, G. B. 1995. *Alaska 1899: Essays from the Harriman Expedition*. Introduction by P. Burroughs and V. Wyatt. Seattle: University of Washington Press.

———. 2007. *The Harriman Expedition to Alaska: Encountering the Tlingit and Eskimo in 1899*. Fairbanks: University of Alaska Press.

Grinnell, J., and T. I. Storer. 1916. "Animal Life as an Asset to National Parks." *Science* 44:375–80.

Gura, P. F. 2007. *American Transcendentalism: A History*. New York: Hill and Wang.

Haber, G., and M. Holleman. 2013. *Among Wolves: Gordon Haber's Insights into Alaska's Most Misunderstood Animal*. Fairbanks: University of Alaska Press.

Harper, F. 1935. "The Name of the Gopher Frog." *Proceedings of the Biological Society of Washington* 48:79–82.

Harvey, M. W. T. 1994. *A Symbol of Wilderness: Echo Park and the American Conservation Movement*. Seattle: University of Washington Press.

Harvey, M. 2005. *Wilderness Forever: Howard Zahniser and the Path to the Wilderness Act*. Seattle: University of Washington Press.

Hayden, F. V. (1872) 2006. *Preliminary Report of the United States Geological Survey of Wyoming and Portions of the Contiguous Territories.* Washington, DC: Elibron Classics.

Hedgpeth, J. W. 1978a. *The Outer Shores. Part 1: Ed Ricketts and John Steinbeck Explore the Pacific Coast.* Eureka, CA: Mad River Press.

———. 1978b. *The Outer Shores. Part 2: Breaking Through.* Eureka, CA: Mad River Press.

Hoagland, S. J. 2016. "A Tribal Model of Wildlife Stewardship: Native Americans Tap into Traditional Practices to Manage Forest." *Wildlife Professional* 10:28–30.

Holovich, A., and K. Reyna. 2016. Review of *The Real Wolf: The Science, Politics, and Economics of Co-Existing with Wolves in Modern Times*, by T. B. Lyons and W. N. Graves. *Journal of Wildlife Management* 80:1332–33.

Hornaday, W. T. 1886. *The Extermination of the American Bison.* Washington, DC: Smithsonian Institution Press.

———. 1929. *Two Years in the Jungle.* New York: Charles Scribner's and Sons.

Houston, C. S. 1988. "In Memoriam: Hans Albert Hochbaum." *The Auk* 105:769–72.

Humboldt, A. von. 1856. *Cosmos: A Sketch or a Physical Description of the Universe.* Translated by E. C. Otté. New York: Harper & Brothers.

Hyman, L., and E. Hutchinson. 1991. "Libbie Hyman, 1888–1969: A Biographical Memoir." *National Academy of Sciences*, pp. 101–14.

Iltis, H. H. 1996. "Alwyn Howard Gentry: A Tribute." *Annals of the Missouri Botanical Garden* 83:446–49.

Irving, W. (1832) 2013. *A Tour on the Prairies: An Account of Thirty Days in Deep Indian Country.* New York: John W. Lovell Company; repr., New York: Skyhorse Publishing.

———. 1836. *Astoria; or, Anecdotes of an Enterprise Beyond the Rocky Mountains.* My copy is an offprint from San Bernardino, California.

Johnsgard, P. A., 1989. Introduction to *Lewis and Clark: Pioneering Naturalists*, by P. R. Cutright. Lincoln: University of Nebraska Press.

Juday, C. 1915. "Limnological Studies on Some Lakes in Central America." *Wisconsin Academy of Sciences, Arts, and Letters* 18:214–50.

Kalmbach, E. R. 1963. "In Memoriam: W. L. McAtee." *The Auk* 80:474–85.

Kingsland, S. E. 2005. *The Evolution of American Ecology, 1890–2000.* Baltimore: Johns Hopkins University Press.

Kirksey, E. 2015. *Emergent Ecologies.* Durham, NC: Duke University Press.

Kline, B. 1997. *First Along the River: A Brief History of the U.S. Environmental Movement.* San Francisco: Acada Books.

Kohler, R. E. 2002. *Landscapes and Labscapes: Exploring the Lab-Field Border in Biology.* Chicago: University of Chicago Press.

———. 2006. *All Creatures: Naturalists, Collectors, and Biodiversity, 1850–1950.* Princeton, NJ: Princeton University Press.

———. 2008. "From Farm and Family to Career Naturalist: The Apprenticeship of Vernon Bailey." *Isis* 99:28–56.

Kohlstedt, S. G. 1985. "Henry Augustus Ward and American Museum Development." *University of Rochester Library Bulletin* 38 (adapted from the 1980 article "Henry A. Ward: The Merchant Naturalist and American Museum Development." *Journal of the Society for the Bibliography of Natural History* [now the *Archives of Natural History*] 9:647–61).

Kurth, J. W. 2014. "The Wilderness Act Turns 50." *Wildlife Professional* 8:11.

Lannoo, M. J. 2008. *Malformed Frogs.* Berkeley: University of California Press.

———. 2010. *Leopold's Shack and Ricketts's Lab: The Emergence of Environmentalism*. Berkeley: University of California Press.

———. 2012. *The Iowa Lakeside Laboratory: A Century of Discovering the Nature of Nature*. Iowa City: University of Iowa Press.

Lendt, D. L. 1989. *Ding: The Life of Jay Norwood Darling*. Ames: Iowa State University Press.

Leopold, A. (1933) 1986. *Game Management*. New York: Charles Scribner's Sons; repr., Madison: University of Wisconsin Press.

Lewis, D. 2012. *The Feathery Tribe: Robert Ridgway and the Modern Study of Birds*. New Haven, CT: Yale University Press.

Lin, M. 2000. *Boundaries*. New York: Simon and Shuster.

Little, J. T. 2000. "A Wilderness Apprenticeship: Olaus Murie in Canada, 1914–1915 and 1917." *Environmental History* 5:531–44.

Logan, B., P. Singleton, C. Thompson, V. Saab, and W. Black. 2016. "Wildfire!" *Wildlife Professional* 10:38–41.

Lorbiecki, M. 1996. *Aldo Leopold: A Fierce Green Fire*. Helena, MT: Falcon Publishing.

Lorimer, J. 2015. *Wildlife in the Anthropocene*. Minneapolis: University of Minnesota Press.

Lyons, T. B., and W. N. Graves, 2014. *The Real Wolf: The Science, Politics, and Economics of Co-Existing with Wolves in Modern Times*. Helena, MT: Farcountry Press.

MacCreagh, G. 1926. *White Waters and Black*. New York: Century Company.

———. 1928. *The Last of Free Africa*. New York: D. Appleton-Century.

Mackovjack, J. 2010. *Tongass Timber: A History of Logging and Timber Utilization in Southeast Alaska*. Durham, NC: Forest History Society.

Maclean, N. 1976. "USFS 1919: The Ranger, the Cook, and a Hole in the Sky." In *A River Runs Through It and Other Stories*. Chicago: University of Chicago Press.

———. 1993. *Young Men and Fire*. Chicago: University of Chicago Press.

Martin, J., and T. M. Franklin. 2012. "A Time for Leadership: Facing a Tsunami of Conservation Challenges." *Wildlife Professional* 6:10–11.

McCabe, R. A. 1987. *Aldo Leopold: The Professor*. Madison, WI: Palmer Publications.

McHugh, T. 1972. *The Time of the Buffalo*. Lincoln: University of Nebraska Press.

McIntosh, R. P. 1975. "H. A. Gleason—'Individualistic Ecologist,' 1882–1976: His Contribution to Ecological Theory." *Bulletin of the Torrey Botanical Club* 102:253–73.

Meine, C. 1988. *Aldo Leopold: His Life and Work*. Madison: University of Wisconsin Press.

Merriam, C. H. 1894. *U.S. Department of Agriculture Report of Ornithologist and Mammalogist for 1983*. Washington, DC: Government Printing Office.

Meyer, J. M. 1997. "Gifford Pinchot, John Muir, and the Boundaries of Politics in American Thought." *Polity* 30:267–84.

Michelant, P. M. H. 1865. *Voyage de Jacques Cartier au Canada en 1534, Nouvelle Edition*. Publiée d'Aprés l'Edition de 1598 et d'Aprés Ramusio, Librairie Tross, Paris.

Millard, C. 2006. *River of Doubt: Theodore Roosevelt's Darkest Journey*. New York: Doubleday.

Miller, C. 2001. *Gifford Pinchot and the Making of Modern Environmentalism*. Washington, DC: Island Press.

Miller, G. S., Jr. 1928. "Mammalogy and the Smithsonian Institution." In *Annual Report of the Board of Regents of the Smithsonian*, pp. 391–411. Washington, DC: Smithsonian Institute.

Minteer, B. A. 2006. *The Landscape of Reform: Civic Pragmatism and Environmental Thought in America*. Cambridge, MA: MIT Press.

Mitchell, L. C. 1981. *Witness to a Vanishing America*. Princeton, NJ: Princeton University Press.

Mittermeier, R. 2011. Foreword to *Still Counting . . . Biodiversity Exploration for Conservation: The First 20 Years of the Rapid Assessment Program*, edited by L. E. Alonso, J. L. Deichmann, S. A. McKenna, P. Naskrecki, and S. J. Richards. Washington, DC: Conservation International.

Moore, J. G. 2006. *King of the 40th Parallel: Discovery on the American West*. Stanford, CA: Stanford University Press.

Mortimer, C. H. 1956. "An Explorer of Lakes." In *E. A. Birge: A Memoir*, by G. C. Sellery, pp. 165–211. Madison: University of Wisconsin Press.

Muir, J. (1912/1913) 1965. *The Story of My Boyhood and Youth*. Madison: University of Wisconsin Press.

———. 1916. *A Thousand-Mile Walk to the Gulf*. Edited by William Frederic Badè. New York: Houghton Mifflin.

Murie, A. 1940. *Ecology of the Coyote in the Yellowstone*. Fauna of the United States, Bulletin Number 4, United States Department of the Interior, National Park Service, Washington, DC.

———. 1944. *The Wolves of Mount McKinley*. Fauna of the National Parks of the United States, Fauna Series Number 5, U.S. Government Printing Office, Washington, DC.

———. 1961. *A Naturalist in Alaska*. New York: Devin-Adair Company.

Murie, M. 1957. *Two in the Far North*. New York: Alfred A. Knopf.

Nash, R. F. 1989. *The Rights of Nature: A History of Environmental Ethics*. Madison: University of Wisconsin Press.

———. 2001. *Wilderness and the American Mind*. 4th ed. New Haven, CT: Yale University Press.

Nicholson, M. 1990. "Henry Allan Gleason and the Individualistic Hypothesis: The Structure of a Botanist's Career." *Botanical Review* 56:91–161.

Nicholson, M., and R. P. McIntosh. 2002. "H. A. Gleason and the Individualistic Hypothesis Revisited." *Bulletin of the Ecological Society of America* 83:133–42.

Nicollet, Joseph N. 1993. *Joseph N. Nicollet on the Plains and Prairies: The Expeditions of 1838–39 with Journals, Letters, and Notes on the Dakota Indians*. Edited by E. C. Bray and M. C. Bray. St. Paul: Minnesota Historical Society Press.

Nijhuis, M. 2014. "Bridging the Conservation Divide." *New Yorker*, December 9. http://www.new yorker.com/tech/elements/bridging-conservation-divide.

Nisbet, J. 2009. *The Collector: David Douglas and the Natural History of the Northwest*. Seattle: Sasquatch Books.

Normile, D. 2016. "El Niño's Warmth Devastating Reefs Worldwide: Recent Aerial Surveys of Australia's Great Barrier Reef Find Massive Coral Bleaching." *Science* 352:15–16.

Nuttall, T. (1819) 1980. *A Journal of Travels into the Arkansas Territory During the Year 1819*. Philadelphia: T. H. Palmer; repr., Norman: University of Oklahoma Press, with an introduction by S. Lottinville.

Oelschlaeger, M. 1991. *The Idea of Wilderness*. New Haven, CT: Yale University Press.

Oliver, T. H. 2016. "How Much Biodiversity Loss Is Too Much?" *Science* 353:220–21.

Olson, S. 1984. *The Singing Wilderness*. Minneapolis: University of Minnesota Press.

Osgood, W. H. 1943. "Clinton Hart Merriam." *Journal of Mammalogy* 24:412–36.

Otteson, J. R. 2014. *What Adam Smith Knew: Moral Lessons on Capitalism from Its Greatest Champions and Fiercest Opponents*. New York: Encounter Books.

Overfield, R. A. 1993. *Science with Practice: Charles E. Bessey and the Maturing of American Botany*. Ames: Iowa State University Press.

Pagnamenta, P. 2012. *Prairie Fever: British Aristocrats in the American West, 1830–1890*. New York: Norton.

Parker, G. H. 1938. "Biographical Memoir of William Morton Wheeler, 1865–1937." *National Academy of Sciences* 19:201–41.

Pauley, P. J. 2000. *Biologists and the Promise of American Life*. Princeton, NJ: Princeton University Press.

Pearson, G. A. 1922. "Preservation of Natural Areas in National Forests." *Ecology* 3:284–87.

Philbrick, N. 2003. *Sea of Glory: America's Voyage of Discovery, the U.S. Exploring Expedition, 1838–1842*. New York: Penguin.

Pick, N., and M. Sloan, with an introduction by E. O. Wilson. 2004. *The Rarest of the Rare: Stories Behind the Treasures at the Harvard Museum of Natural History*. New York: HarperCollins.

Pinchot, G. (1947) 1998. *Breaking New Ground*. First edition published privately by the estate of Gifford Pinchot. My copy is the 1998 edition published by Island Press, Washington, DC, with an introductory essay by Char Miller and V. Alaric Sample.

Powell, J. W. 2013. *The Exploration of the Colorado River and Its Canyons*. Hollywood, FL: Simon and Brown.

Preston, R. 2007. *The Wild Trees: A Story of Passion and Daring*. New York: Random House.

Pritchard, J. A., D. M. Debinski, B. Olechnowski, and R. Vannimwegen. 2006. "The Landscape of Paul Errington's Work." *Wildlife Society Bulletin* 34:1411–16.

Rebok, S. 2014. *Humboldt and Jefferson: A Transatlantic Friendship of the Enlightenment*. Charlottesville: University of Virginia Press.

Reeves, H. M., and D. B. Marshall. 1985. "In Memoriam: Ira Noel Gabrielson." *The Auk* 102: 865–68.

Remsen, J. V., Jr., and T. S. Schulenberg. 1997. "The Pervasive Influence of Ted Parker on Neotropical Field Ornithology." *Ornithological Monographs* 48:7–19.

Ricketts, E. F. 2006. *Breaking Through: Essays, Journal, and Travelogues of Edward F. Ricketts*. Edited by K. A. Rodger. Berkeley: University of California Press.

Ricklefs, R. F., and G. L. Miller. 2000. *Ecology*. 4th ed. New York: W. H. Freeman.

Ripley, S. D., and J. A. Steed. 1987. "Alexander Wetmore, 1886–1978: A Biographical Memoir." *National Academy of Sciences*, pp. 597–626.

Rivinus, E. F., and E. M. Youssef. 1992. *Spencer Baird of the Smithsonian*. Washington, DC: Smithsonian Institution Press.

Robinson, M. J. 2005. *Predatory Bureaucracy: The Extermination of Wolves and the Transformation of the West*. Boulder: University Press of Colorado.

Rodger, K. A. 2002. *Renaissance Man of Cannery Row*. Tuscaloosa: University of Alabama Press.

Rodgers A. D., III. 1944. *John Merle Coulter: Missionary in Science*. Princeton, NJ: Princeton University Press.

———. 1949. *Liberty Hyde Bailey: A Story of American Plant Sciences*. Princeton, NJ: Princeton University Press.

———. 1965. *John Torrey: A Story of North American Botany*. New York: Hafner Press.

Roosevelt, T. (1910) 2015. *Biological Analogies in History*. London: Oxford University Press; repr., Perfect Library.

———. (1920) 2009. *The Autobiography of Theodore Roosevelt*. Repr., Seven Treasures Publications.

Royte, E. 2008. "Night Moves." *New York Times Book Review*, June 22.

Russell, H. L. 1940. "Dr. Birge as a Teacher." In *Edward A. Birge: Teacher and Scientist*, Symposium on Hydrobiology, pp. 7–14. Madison: University of Wisconsin Press.

Sagarin, R., and A. Pauchard. 2012. *Observation and Ecology: Broadening the Scope of Science to Understand a Complex World*. Washington, DC: Island Press.

Savage, H., Jr. 1979. *Discovering America, 1700–1875*. New York: Harper & Row.

Schaller, G. B. 1980. *Stones of Silence: Journeys in the Himalaya*. Budapest, Hungary: André Deutsch.

Schlachtmeyer, S. S. 2010. *A Death Decoded: Robert Kennicott and the Alaska Telegraph*. Alexandria, VA: Voyage Publishing.

Schmidly, D. J., W. E. Tydeman, and A. L. Gardner, eds. 2016. *United States Biological Survey: A Compendium of its History, Personalities, Impacts, and Conflicts*. Special Publication Number 64. Lubbock: Museum of Texas Tech University.

Schmidt, K. P. 1957. "Warder Clyde Allee, 1885–1955: A Biographical Memoir." *National Academy of Sciences*, pp. 1–40.

Seager, S. 2016. "How Can We Build Another Earth," CNN.com, October 7, http://www.cnn.com /2016/10/07/opinions/sara-seager-exoplanets-terraforming.

Sears, P. B. 1964. *Biology of the Living Landscape*. Sydney, Australia: Allen and Unwin.

Sellery, G. C. 1956. *E. A. Birge: A Memoir*. Madison: University of Wisconsin Press.

Shell, H. R. 2004. "Skin Deep: Taxidermy, Embodiment and Extinction in W. T. Hornaday's Buffalo Group." *Proceedings of the California Academy of Sciences*. Supplement I, no. 5:88–112.

Shepard, O., ed. (1927) 1961. *The Heart of Thoreau's Journals*. New York: Dover.

Shoaf, B. 2000. *The Taking of the Tongass: Alaska's Rainforest*. Sequim, WA: Running Wolf Press.

Sivils, M. 2012. "Introduction: Paul L. Errington: His Life and Work." In *Of Men and Marshes*. Iowa City: University of Iowa Press.

Soderberg, K. A., and J. DuRette. 1988. *People of the Tongass: Alaska Forestry under Attack*. Bellevue, WA: Free Enterprise Press.

Souder, W. 2000. *A Plague of Frogs*. New York: Hyperion Press.

Sperry, T. M. 1935. "Root Systems in Illinois Prairie." *Ecology* 16:178–202.

Stark, P. 2014. *Astoria: Astor and Jefferson's Lost Empire: A Tale of Ambition and Survival on the Early American Frontier*. New York: Ecco Press.

Stegner, W. (1954) 1992. *Beyond the Hundredth Meridian: John Wesley Powell and the Second Opening of the American West*. New York: Houghton Mifflin; repr., New York: Penguin.

———. (1971) 2014. *Angle of Repose*. New York: Vintage Books.

———. 1999. *Marking the Sparrow's Fall: The Making of the American West*. Edited by P. Stegner. New York: Henry Holt.

Stein, B. R. 2001. *On Her Own Terms: Annie Montague Alexander and the Rise of Science in the American West*. Berkeley: University of California Press.

Steinbeck, E., and R. Wallsten, eds. 1989. *Steinbeck: A Life in Letters*. London: Penguin.

Steinbeck, J. 1939. *Grapes of Wrath*. New York: Viking Press.

Steinbeck, J., and E. F. Ricketts. (1941) 1971. *Sea of Cortez: A Leisurely Journal of Travel and Research*. New York: P. P. Appel.

Sterling, K. B. 1977. *Last of the Naturalists: The Career of C. Hart Merriam*. Rev. ed. New York: Arno Press.

———. 1997a. "Alexander Mackenzie." In *Biographical Dictionary of American and Canadian Naturalists and Environmentalists*, edited by K. B. Sterling, R. P. Harmond, G. A. Cevasco, and L. F. Hammond, pp. 490–93. Westport, CT: Greenwood Press.

———. 1997b. "José Mariano Mociño." In *Biographical Dictionary of American and Canadian Naturalists and Environmentalists*, edited by K. B. Sterling, R. P. Harmond, G. A. Cevasco, and L. F. Hammond, pp. 549–52. Westport, CT: Greenwood Press.

Stoddard, H. L., Sr. 1969. *Memoirs of a Naturalist*. Norman: University of Oklahoma Press.

Stoddard, H. L., and P. Errington. 1938. "Modifications in Predation Theory Suggested by Ecological Studies of the Bobwhite Quail." *Transactions of the Third North American Wildlife Conference*. Washington, DC: American Wildlife Institute, pp. 736–40.

Sullivan, R. 1993. "Theodore Parker, Alwyn Gentry, Biologists, Die in Airplane Crash." *New York Times*, August 6, p. B7.

Sutter, P. S. 2002. *Driven Wild: How the Fight against Automobiles Launched the Modern Wilderness Movement*. Seattle: University of Washington Press.

Symposium on Hydrobiology. 1940. *Edward A. Birge: Teacher and Scientist*. Madison: University of Wisconsin Press.

Tamm, E. E. 2004. *Beyond the Outer Shores: The Untold Odyssey of Ed Ricketts, the Pioneering Ecologist Who Inspired John Steinbeck and Joseph Campbell*. New York: Thunder's Mouth Press.

Tanner, T., ed. 1987. *Aldo Leopold: The Man and His Legacy*. Ankeny, IA: Soil Conservation Society.

Teale, E. W. (1954) 2001. *The Wilderness World of John Muir*. New York: Houghton Mifflin; repr., New York: Mariner Books.

Thomas, J. W. 2015. *Forks in the Trail: A Conservationist's Trek to the Pinnacles of Natural Resource Leadership*. Missoula, MT: Boone and Crockett Club.

Tidball, E. C. 2004. *Soldier-Artist of the Great Reconnaissance: John C. Tidball and the 35th Parallel Pacific Railroad Survey*. Tucson: University of Arizona Press.

Tobey, R. C. 1981. *Saving the Prairies: The Life Cycle of the Founding School of American Plant Ecology, 1895–1955*. Berkeley: University of California Press.

Townsend, J. K. (1839) 1999. *Narrative of a Journey across the Rocky Mountains to the Columbia River*. Edited by G. A. Jobanck. Corvallis: Oregon State University Press.

Turner, F. J. 1935. *The Frontier in American History*. New York: Henry Holt.

Turrill, W. B. 1963. *British Men of Science: Joseph Dalton Hooker*. London: Thomas Nelson and Sons.

Vitt, L. J. 2013. "Walking the Natural-History Trail." *Herpetologica* 69:105–17.

Voigt, J. W. 1980. "J. E. Weaver and the North American Prairie: 'Look Carefully and Look Often.'" From an address delivered at the Seventh North American Prairie Conference, Southwestern Missouri State University, Springfield, Missouri, on August 4, p. 320. Available online at http://images.library.wisc.edu/EcoNatRes/EFacs/NAPC/NAPC07/reference/econat res.napc07.jvoigt.pdf.

Van Nuys, F. 2015. *Varmints and Victims: Predator Control in the American West*. Lawrence: University of Kansas Press.

Wallace, J. 2000. *A Gathering of Wonders: Behind the Scenes at the American Museum of Natural History*. New York: St. Martin's Press.

Washburn, W. E. 1978. *The Cosmos Club of Washington: A Centennial History*. Washington, DC: Cosmos Club.

Weaver, J. E. 1954. *North American Prairie*. Chicago: Johnsen Publishing.

Weaver, J. E., and F. E. Clements. 1938. *Plant Ecology*. New York: McGraw-Hill.

Weinberg, S. 2000. *A Fish Caught in Time*. New York: Harper Perennial.

Welch, M. 1998. *The Book of Nature: Natural History in the United States, 1825–1875*. Boston: Northeastern University Press.

Welch, P. S. 1944. "Chancey Juday." *Ecology* 25:271–72.

Wetmore, F. A. 1915. "Mortality among Waterfowl around Great Salt Lake, Utah (Preliminary Report)." *United States Department of Agriculture Bulletin* 217:1–10.

———. 1918. "The Duck Sickness in Utah." *United States Department of Agriculture Bulletin* 672:1–26.

———. 1919. "Lead Poisoning in Waterfowl." *United States Department of Agriculture Bulletin* 793:1–12.

———. 1926. *The Migrations of Birds.* Cambridge, MA: Harvard University Press.

Williams, T. T. 2016. *The Hour of Land: A Personal Topography of America's National Parks.* New York: Sarah Crichton Books, Simon and Shuster.

Wilson, E. O. 1994. *Naturalist.* Washington, DC: Island Press.

———. 2004. "Introduction." In *The Rarest of the Rare: Stories Behind the Treasures at the Harvard Museum of Natural History,* by N. Pick and M. Sloan. New York: HarperCollins.

Windsor, M. P. 1991. *Reading the Shape of Nature: Comparative Zoology at the Agassiz Museum.* Chicago: University of Chicago Press.

Witte, S. S., and M. V. Gallagher, eds. 2008. *The North American Journals of Prince Maximilian of Wied,* Volume II, *April–September 1833.* Translated by W. J. Orr, P. Schach, and D. Karch. Norman: University of Oklahoma Press.

Wulf, A. 2015. *The Invention of Nature: Alexander von Humboldt's New World.* New York: Knopf.

Wydeven, A. P. 2016. Review of *The Real Wolf: The Science, Politics, and Economics of Co-Existing with Wolves in Modern Times,* by T. B. Lyons and W. N. Graves [Helena, MT: Farcountry Press, 2014]. *Journal of Wildlife Management* 80:1334–35.

Index